Design of Systems on a Chip: Design and Test

Edited by

Ricardo Reis
Federal University of Rio Grande do Sul

Marcelo Lubaszewski
Federal University of Rio Grande do Sul

and

Jochen A.G. Jess
Eidhoven University of Technology

 Springer

A C.I.P. Catalogue record for this book is available from the Library of Congress.

ISBN-10 0-387-32499-2 (HB)
ISBN-13 978-0-387-32499-9 (HB)
ISBN-10 0-387-32500-X (e-book)
ISBN-13 978-0-387-32500-2 (e-book)

Published by Springer,
P.O. Box 17, 3300 AA Dordrecht, The Netherlands.

www.springer.com

Printed on acid-free paper

CONTENTS

CONTRIBUTORS

Prof. Anceau received the engineering degree from the *Institut National Polytechnique de Grenoble* (INPG) in 1967, and the *Docteur d'Etat* degree in 1974 from the University of Grenoble. He started his research activity in 1967 as member of the *Comite National pour la Recherche Scientifique* (CNRS) in Grenoble. He became Professor at INPG in 1974 where he led a research team on microprocessor architecture and VLSI design. In 1984, he moved to industry (BULL company, close to Paris) to lead a research team on Formal Verification for Hardware and Software. The first industrial tool for hardware formal verification and the technique of symbolic state traversal for finite state machines was developed in this group. In 1996 he took his present position as Professor at *Conservatoire National des Arts et Metiers* (CNAM) in Paris. Since 1991 he has also been a Professor at *Ecole Polytechnique* in Palaiseau, France. His main domains of interest are: microprocessor architecture, VLSI design, hardware formal verification, and research management and strategy. He has given many talks on these subjects. He his the author of many conference papers and of a book entitled "The Architecture of Microprocessors" published by Addison-Wesley in 1986. He launched the French Multi-Chip-Project, called CMP, in 1981.

Dr. Don Bouldin is Professor of Electrical & Computer Engineering at the University of Tennessee. He has served as an instructor for eight NSF-sponsored faculty enhancement short courses on synthesis and FPGAs and presented seminars on this subject in India, Canada, Thailand, Argentina, Brazil and South Korea. He was the 1995-96 Editor-in-Chief of the *IEEE Transactions on VLSI Systems* and Technical Program Chairman for ISCAS-96. Dr. Bouldin was named a Fellow of the IEEE for contributions to the design of special-purpose architectures using VLSI processors. He has authored more than 230 publications and been the Principal Investigator for over nine million dollars of sponsored research.

Ahmed Amine Jerraya (A'94) received the Engineer degree from University of Tunis in 1980 and the D.E.A., "Docteur Ingénieur" and the "Docteur d'Etat" degrees from the University of Grenoble in 1981, 1983 and 1989, respectively, all in computer sciences.

In 1980 he became a Member of the Staff of the Laboratory IMAG in Grenoble, working on CAD for VLSI. In 1986, he held a full research position with CNRS (Centre National de la Recherche Scientifique). From April 1990 to March 1991, he was a Member of the Scientific Staff at Bell-Northern Research Ltd., in Canada, working on linking system design tools and hardware design environments. In 1991, he returned to the University of Grenoble, in the Laboratory TIMA (Techniques of Informatics and Microelectronics) where he founded the system-level synthesis group. He received the Award of President de la Republique in Tunisia, 1980, the highest computer science engineer degree. He participated to LUCIE system, which are highly successful layout tools, distributed in the early 1980s. Since 1992, he has headed the AMICAL project, a highly successful architectural synthesis tool.

Dr. Jerraya served as Technical Program and Organization Chair of the International Symposium on System Synthesis, as Program Chair of the 7th Rapid System Prototyping Workshop, as General Chair of the 9th International Symposium on System Synthesis, and in the program committee of several international conferences including ICCAD, ICCD, ED&TC, EuroDAC and EuroVHDL. He received the Best Paper Award at the 1994 ED&TC for his work on hardware/software co-simulation. He also served as the Program Co-Chair of Codes/CASHE 1998, the general Co-Chair of CODES'99 and he is the Vice-General Chair of DATE 2000.

Jochen A.G. Jess graduated in 1961 for his Masters and in 1963 for his Ph. D. degree in Electrical Engineering at Aachen University of Technology. In 1971 he was appointed as a full professor with tenure to establish and run the Design Automation Section of Eindhoven University of Technology. In the almost 30 years at Eindhoven he himself and his group contributed to almost any relevant subject in the area of VLSI chip design. Until his retirement from the university in the year 2000 the group published some 380 papers. Under his supervision 45 students graduated for their Ph. D. degree. He is one of the founders of the "Design Automation and Test in Europe" (DATE) conference. After his retirement he started as a consultant for IBM Th. J. Watson Research Center and the Embedded Systems Institute of Eindhoven University. Currently he is a consultant to Philips Research at Eindhoven. In addition he teaches courses for professionals in chip architecture for the Center of Technical Training (CTT) of Philips and for the Embedded Systems Institute (ESI) of Eindhoven University. In 2005 he received the EDAA "Lifetime Achievement Award".

Hans G. Kerkhoff received his M.Sc. degree in Telecommunication with honours at the Technical University of Delft in 1977 and the Delft Hogeschool Award for his M.Sc. thesis. In the same year he became staff member of the chair IC-Technology & Electronics at the Faculty of Electrical Engineering, University of Twente. He obtained a Ph.D. in Technical Science (micro-electronics) at the University of Twente in 1984. In the same year he was appointed associate professor in Testable Design and Testing at the Faculty of Electrical Engineering at the University of Twente. In 1991, he became head of the group "Testable Design and Test of Microsystems" of the MESA Research Institute and of the MESA Test Centre. In 1992 he spend his sabbatical year at the test company Advantest in San Jose, USA. Since 1995, he works in addition part-time at the Philips Research Laboratories at Eindhoven. His interest is in testable design and testing of microsystems.

He advised 8 Ph.D. students in this area and has (co-)authored over 80 publications.

Marcelo Lubaszewski received the Electrical Engineering and the M.Sc. degrees from Universidade Federal do Rio Grande do Sul (UFRGS) – Brazil, in 1986 and 1990 respectively. In 1994, he received the Ph.D. degree from Institut National Polytechnique de Grenoble (INPG) – France. In 2001 and 2004, he was in sabbatical respectively at Laboratoire d'Informatique, Robotique et Microelectronique de Montpellier (LIRMM) – France, and at Instituto de Microelectrónica de Sevilla (IMSE) – Spain. Since 1990 he is a Professor at UFRGS, where he lectures on microprocessor-based systems and on VLSI design and test. His primary research interests include design and test of mixed-signal, microelectro-mechanical and core-based systems, self-checking and fault-tolerant systems, and computer-aided testing. He has published over 150 papers on these topics. Dr. Lubaszewski has served as the general chair or program chair of several conferences. He has also served as a Guest Editor of the KAP Journal of Electronic Testing: Theory and Applications and as an Associate Editor of the IEEE Design and Test of Computers Magazine.

Ricardo Reis is a professor at the *Instituto de Informática* of the *Universidade Federal do Rio Grande do Sul*. Reis has a BSc in electrical engineering from the *Universidade Federal do Rio Grande do Sul*, and a PhD in computer science, option microelectronics from the Institut National Polytechnique de Grenoble, France. His research interests include VLSI design, CAD, physical design, design methodologies, and fault-tolerant techniques. He has more than 300 publications in journals, conferences and books. He is trustee of the International Federation for Information Processing; trustee of the Brazilian Microelectronics Society and also of the Brazilian Computer Society. Prof. Reis has served as the general chair or program chair of several conferences. He is the Latin America liaison for IEEE Design & Test. He is also editor-in-chief of the Journal of Integrated Circuits and Systems, JICS.

Satnam Singh is interested in developing tools and techniques to exploit dynamic reconfiguration and to find new applications for FPGA-based accelerator cards. From 1991 to 1997 he worked at the University of Glasgow where he managed FPGA projects e.g. high speed PostScript rendering and the application of partial evaluation for the dynamic specialisation of XC6200 circuits. He joined Xilinx in January 1998. He has also done work for European Silicon Structures, Compass Design Automation and British Telecom. He received his PhD from the University of Glasgow in 1991 for research in the area of formal techniques for the systematic development of VLSI CAD.

Dr Wilkinson holds a Degree in Physics from Oxford University, an MSc in Microelectronics from Southampton University and a Doctorate for research into semiconductor devices. Following eight years in the semiconductor industry, he joined Burroughs Corporation, U.S.A., becoming a key contributor to the formulation of technology strategy and its implementation across the USA and the UK.

In 1985 he returned to Cambridge to work for Sinclair Research. He was one of the founders of Anamartic Limited, a memory products design and manufacturing company.

In 1992 Dr Wilkinson founded TFI Limited, a technology transfer and management consultancy specialising in microsystems.

Zorian is the vice president and chief scientist of Virage Logic Corporation and an adjunct professor at the University of British Columbia. Previously a Distinguished Member of Technical Staff at AT&T Bell Laboratories and chief technology advisor of LogicVision. Dr. Zorian received an MSc from the University of Southern California and a PhD from McGill University. He served as the IEEE Computer Society Vice President for Conferences and Tutorials, Vice President for Technical Activities, Chair of the IEEE Test Technology Technical Council, and editor in chief of IEEE Design & Test of Computers. He has authored over 300 papers, holds 13 US patents, and received numerous best-paper awards and Bell Labs' R&D Achievement Award. An IEEE Fellow, he was selected by EE Times among the top 13 influencers on the semiconductor industry. Dr. Zorian was the 2005 recipient of the prestigious IEEE Industrial Pioneer Award.

CHAPTER 1

DESIGN OF SYSTEMS ON A CHIP
Introduction

MARCELO LUBASZEWSKI[1], RICARDO REIS[2], AND JOCHEN A.G. JESS[3]

[1] *Electrical Engineering Department, Universidade Federal do Rio Grande do Sul (UFRGS), Av. Osvaldo Aranha esquina Sarmento Leite 103, 90035-190 Porto Alegre RS, Brazil;*
[2] *Institute of Computer Science, Universidade Federal do Rio Grande do Sul (UFRGS), Av. Bento Gonçalves 9500, Cx. Postal 15064, 91501-970 Porto Alegre RS, Brazil;*
[3] *Eindhoven University of Technology, p.o.box 513, 5600 MB Eindhoven, The Netherlands, Phone: 31-40-247-3353, Fax 31-40-246-4527*

Abstract: This introductory chapter briefly discusses the impacts into chip design and production of integrating highly complex electronic systems as systems on a chip. Technology, productivity and quality are the main aspects under consideration to establish the major requirements for the design and test of upcoming systems on a chip. The contents of the book are shortly presented in the sequel, comprising contributions on three different, but complementary axes: core design, computer-aided design tools and test methods

Keywords: Microelectronics, integrated circuits, VLSI, systems on a chip, cored-based systems, CAD, VLSI testing, design-for-test

1. CHIP COMPLEXITY: IMPACT INTO DESIGN

Chips are nowadays part of every equipment, system or application that embeds electronics. Contrarily to the first decades that followed the transistor invention, the semiconductors market is presently one of the most important segments of the world economy and has become strategic for any country that plans to get some technology independence. In fact, this technology context tends to prevail in next years, since the electronic world will continue expanding fastly through the internet and the wireless communication.

Currently, in a single chip of tenths of square milimeters it is possible to integrate hundreds of thousands of transistors and hundreds of passive components. These chips are real integrated systems and, because of that, they are called systems on a chip. The performance of these systems reaches few gigahertz, while the power

1

R. Reis et al. (eds.), Design of Systems on a Chip, 1–7.
© 2006 *Springer.*

consumption is of the order of milliwatts. In fact, the growth of the integration density in microelectronics technologies has followed very closely the Moore's Law announced in the sixties: every year the integration density will double, resulting in an exponential growth along the years. Additionally to the increasing electronics density, technology advances have allowed to integrate heterogeneous parts on the same substrate. Digital, analog, thermal, mechanical, optical, fluidic and other esoteric functions can now be put together into the same chip (ITRS, 1999; ITRS, 2003). This obviously adds to chip complexity.

In terms of design, the price to pay for increasing integration scales, densities, performances, functionality, and decreasing sizes and power consumption, is an important growth in the complexity of the design flow. This cost roughly translates into spending more time to produce a new design, or hiring more skilled designers.

In respect to time-to-market, considering the real-life competitiveness, the later a product arrives to market, the lower are the revenues got from. This is illustrated in the simplified model given in Figure 1 (Mourad, 2000).

Starting from the arrival to market, the revenues grow to a peak and then start decreasing to the end of the product cycle. The polygon surface is the total revenues that a product can give along time. A ΔT delay in time-to-market will transfer part of the revenues to competitors, and the maximum instant revenue achievable will thus be decreased by the amount transferred at the delayed arrival time. As shown in the figure, this will drastically reduce the polygon surface and will cause an important loss of revenues.

Regarding the design team, since costs cannot exceed product sales, ways of increasing designer's productivity are preferred to hiring ever more people. Figure 2 reports an increasing gap between circuits complexity and designers productivity (ITRS, 1999).

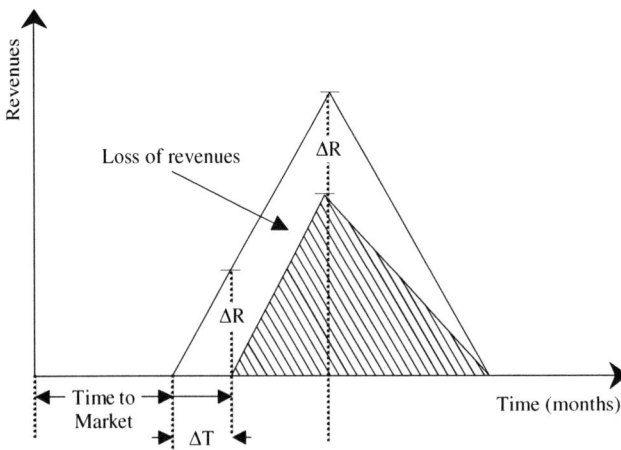

Figure 1. Consequences of time-to-market delays into product revenues

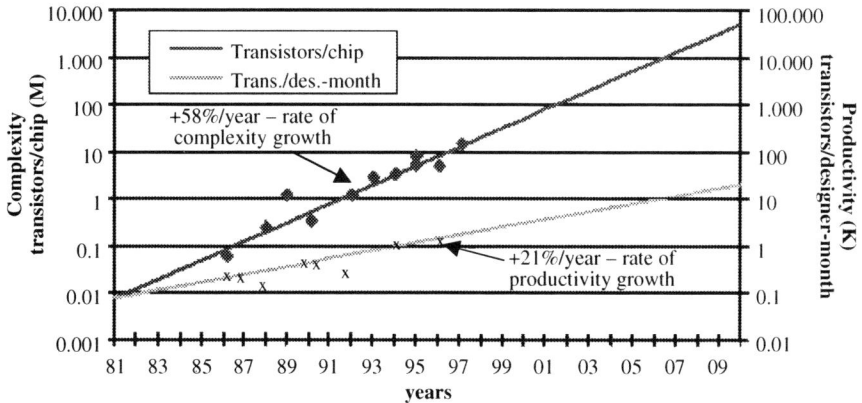

Figure 2. Increasing gap between IC complexity and productivity (ITRS, 1999)

The curves show different growth rates for the number of transistors per integrated circuit and for the number of transistors produced per month by a designer. The former points out to a growth of 58% a year, while the latter only to 21% a year. Reuse and interface standardization initiatives may alleviate the present productivity deficit, but still need to rely on highly automated design environments to cope with the problem.

Typically, state-of-the-art integrated systems are not designed from scratch, designers productivity is enhanced by maximizing reuse. Systems on a chip are currently built from existing microprocessor, memory, programmable logic, mixed-signal, sensor and other pre-designed and pre-verified cores. An appropriate choice of cores, some interfacing work and some user-defined logic design should be enough to customize an integrated system to a given application. However, this system integration work is no simple task. First of all, the selected cores may be described at quite different levels of abstraction: layout, transistor/logic gate netlist, RT level, behavioural HD, etc. Secondly, tools for hardware-software partitioning, for co-simulation, for software compilation, for behavioural, logic and physical hardware synthesis, for design verification and for rapid prototyping shall be available as part of a highly integrated computer-aided design environment to make design possible. Additionally, these CAD tools shall be capable of dealing with hundreds of thousands to a few millions of transistors in an affordable computing time. This can only be feasible if the design environment is highly flexible (accepts descriptions of different levels of abstraction), is highly hierarchical (breaks complexity), allows for step-by-step development (adding details from the system specification to the layout) and makes it possible that detailed design partitions can be dealt with separately and concurrently (improves productivity).

Therefore, design solutions capable to face today's chips complexity shall build around designers with heterogeneous skills (to cope with new technologies), that shall work around core-based systems (to maximize design reuse) and make use of powerful CAD tools (to improve team productivity). Get to this ideal picture is the challenge that has been faced in last years by many educators, researchers and industrials all over the world.

2.   CHIP COMPLEXITY: IMPACT INTO PRODUCTION

Once a new design is ready to go into production, the major features that impact the final fabrication costs are: the integration scale, the transistor density and the chip heterogeneity. They all together determine how complex the integration process, the chip wiring, packaging and testing will be.

To some extent, reuse is also a keyword to make the fabrication of complex chips economically viable. CMOS is definitely the most used technology world-wide. Existing CMOS processes start adding few fabrication steps to allow for the implementation of other than digital functions into the same chip. Analog, optical, mechanical and other functions will ever more join digital, and lead to highly heterogeneous systems on a chip. For specific applications, such as RF, these integrated systems will additionally require special wiring and packaging technologies, increasing even more the chip complexity from the manufacturing point of view.

For a given semiconductor technology, real life chip implementations will be affected by physical defects whose density is a function of the process maturity and the number of fabrication steps. The defects density and the wafer area are the major aspects that determine the fraction of the manufactured chips that will be defect-free. This fraction is known as the IC manufacturing yield. Since the wafer area is increasing constantly, to improve the yield, manufacturers are required, first of all, to understand the existing types of defects. Then, in order to minimize the occurrence of these defects, they need to interfere in the integrated circuit fabrication by improving the various steps of the manufacturing process.

Once the best is obtained from the yield point of view, product reliability can only be further improved by reducing the defect level, that is, the fraction of bad parts that pass the production tests. In practice, low defect levels can only be obtained through tests that provide high fault coverage.

In general, developing and applying these tests take long times and may delay the introduction of the product into the market. Thus, according to the time-to-market model of Figure 1, if it causes delays, and also incurs cost, it turns hard to justify testing on a purely financial basis. However, one should consider that this model does not account for the loss of revenues due to rejects because of problems in the final product. Figure 3 proposes a more realistic model, where product quality, represented by the slopes q and Q, is added to the model of Figure 1: the better the product quality, the greater the slope angle. Then, one can notice from the figure that, although a slight delay may come out due testing reasons, the loss of revenues may be largely compensated for better quality.

As traditional methods, based on external functional tests, continue to be used for increasingly complex systems on a chip, it becomes ever more evident how hard it turns to get good market shares through time and product quality tradeoffs.

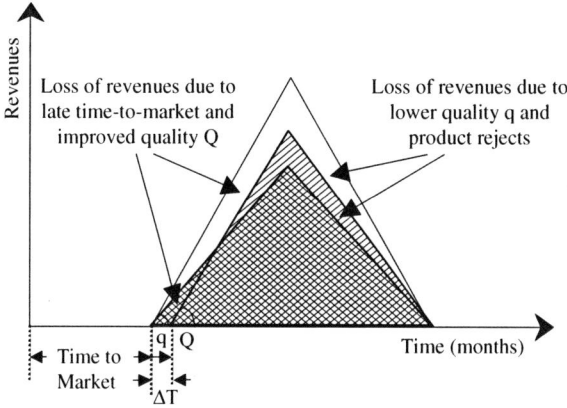

Figure 3. Consequences of rejects into product revenues

State-of-the-art chips have a much larger number of transistors per pin, what makes it much more difficult the access and control of internal nodes by an external tester. This limited access and control leads to a substantial increase in test application times.

Additionally, since these chips operate at much higher frequencies, at-speed testing may only be feasible by using extremely expensive testers that, according to (ITRS, 1999), will cost more than US$20M by the year 2014. The curves in Figure 4 show up how this situation will probably evolve if traditional methods continue dominating the testing scenario: the costs for transistors fabrication will continue falling down and the costs for transistors testing will grow up along the

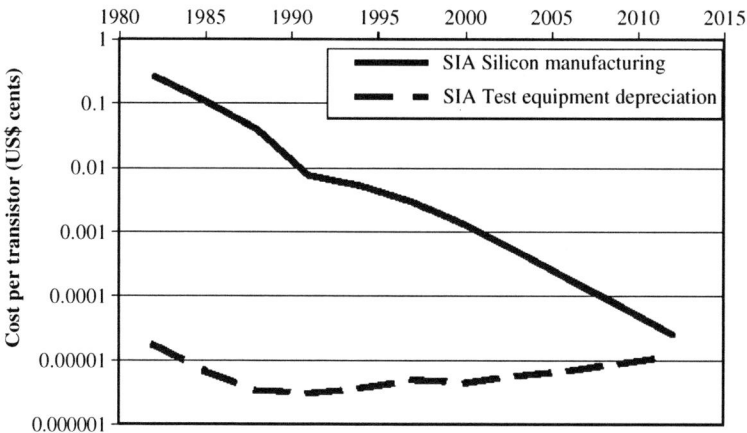

Figure 4. Cost per transistor in microprocessors: fabrication versus testing (ITRS, 1999)

next years (ITRS, 1999; ITRS, 2003). This means that, by the year 2012, testing will start dominating production costs and product competitiveness will establish around cheaper test techniques yet capable of ensuring product quality.

Design-for-test (DfT), that implies moving part or all tester functionalities into the chip, points out as the only viable solution to reduce the tester capability requirements and therefore, the cost of the manufacturing test equipment. Additionally, the use of structured and hierarchical DfT approaches, that provide the means to partition the system and to test multiple parts concurrently, may enable higher fault coverage and shorter test application times. The impact of chip complexity into production may thus be alleviated by DfT, as it ensures product quality at lower costs and simultaneously prevents test from dominating time-to-market (ITRS, 2003).

3.   THE SECOND BOOK: DESIGN AND TEST

This book is the second of two volumes addressing design challenges associated with new generations of the semiconductor technology. The various chapters are the compilation of tutorials presented by prominent authors from all over the world at workshops in Brazil in the recent years. In particular, this second book deals with core design, CAD tools and test methods.

To start with cores, a collection of contributed chapters deal with the heterogeneity aspect of designs, showing the diversity of parts that may share the same substrate in a state-of-the-art system on a chip. Three chapters address mixed-signal cores, in particular microsystems embedding electrical and non-electrical parts, and purely digital cores, such as multimedia processors, conventional microprocessors and reconfigurable logic. In Chapter 2, J. M. Wilkinson discusses the evolution, types, fabrication and application of modern microsystems, including examples of technological strategies suitable for small and medium size high-tech companies. Chapter 3, by J. A. G. Jess, reviews the relevant features of digital media applications, discusses core architectures for multimedia processing and addresses the question of software design for these architectures. In Chapter 4, F. Anceau presents the extraordinary evolution of microprocessors to our days, and discusses the challenging future of these devices to keep the evolutionary rate of computing power and binary code compatibility.

Then it comes the turn of computer-aided design tools. The second part of this book discusses CAD in three very different levels of design abstraction. Chapter 5, by R. Reis, J. Güntzel and M. Johann, looks at the various requirements that physical synthesis tools have to meet, such as layouts with smaller silicon areas, shorter signal delays and lower power consumptions, for implementation in submicron technologies. R. A. Bergamaschi, in Chapter 6, tackles the problem of designer's productivity by proposing synthesis algorithms for behavioural descriptions, and by discussing the main advantages and drawbacks of behavioural synthesis along with its insertion into the overall digital design methodology. Finally, in Chapter 7 A. Jerraya and co-authors discuss how, starting from the system specification level, a heterogeneous architecture composed of software, hardware and communication modules can be generated by means of a co-design tool.

The third and last part of this book, composed of four chapters, deals with test methods. In this collection of chapters, the topic is addressed from various different viewpoints: in terms of chip complexity, test is discussed from the core and system prospective; in terms of signal heterogeneity, the digital, mixed-signal and microsystem prospective are considered. The fundamentals and an overview of the state-of-the-art test and design-for-test methods are given in Chapter 8 by M. Lubaszewski. In Chapter 9, D. W. Bouldin discusses the need, while migrating from a FPGA prototype to an ASIC implementation, to integrate on-chip design-for-test methods that enable to screen out manufacturing defects. H. G. Kerkoff addresses in Chapter 10 test reuse in microsystems and test innovations brought in by the embedded sensors and actuators. Y. Zorian wraps up this part of the book by presenting, in Chapter 11, the challenges of testing core-based system-chips and the corresponding testing solutions for the individual cores and the complete system.

REFERENCES

ITRS, 1999, *The International Technology Roadmap for Semiconductors*; http://public.itrs.net/files/1999_SIA_RoadMap/

ITRS, 2003, *The International Technology Roadmap for Semiconductors*, http://public.itrs.net/

Mourad, S. and Zorian, Y., 2000, *Principles of Testing Electronic Systems*, John Wiley and Sons.

CHAPTER 2

MICROSYSTEMS TECHNOLOGY AND APPLICATIONS

J. MALCOLM WILKINSON

Technology For Industry Limited, 6 Hinton Way, Wilburton, Ely Cambs. CB6 3SE, U.K.,
Tel: +44 1353 741 331 Fax: +44 1353 740 665, e-mail: tfi@dial.pipex.com

Abstract: Microsystems are intelligent miniaturised systems comprising sensing, processing and/or actuating functions, normally combining two or more electrical, mechanical or other properties on a single chip or multi-chip hybrid. They provide increased functionality, improved performance and reduced system cost to a large number of products. This tutorial will provide an overview of the evolution, types, fabrication and application of modern microsystems including examples of technological strategies suitable for direct utilisation by small and medium size high-tech companies. The tutorial is divided into the following sections: Fabrication, Packaging Technology, Design and Test, Applications for Microsystems, Factors Limiting Exploitation, Business Strategies for Microsystems Companies

1.   INTRODUCTION

The world has been changed dramatically by the integrated circuit, which is now the heart of so much electronic equipment. In the 1980s an important development took place when the same technology, which had been used to make integrated circuits, started to be used to make sensing devices and mechanical components from the same starting material – silicon. During the next ten years, pressure sensors, temperature sensors, imaging devices and a variety of more sophisticated sensors were developed.

The first steps were also taken towards the fabrication of actuators such as valves, pumps and moving mirrors. By the early 1990s it was clear that by integrating sensors, signal processing electronics and actuators together, a complete MICROSYSTEM could be formed on one silicon chip.

The benefits of microsystems stem from the use of the same batch fabrication manufacturing methods which have evolved so efficiently to produce conventional integrated circuits. The technology allows thousands of identical devices to be fabricated at the same time with mechanical tolerances at the micron level and almost identical electrical properties. The batch fabrication processes promise lower cost,

9

R. Reis et al. (eds.), Design of Systems on a Chip, 9–25.
© 2006 *Springer.*

better performance, smaller size and improved reliability compared with discrete
sensors fabricated with technology which pre-dates the microchip era.

Another interesting development has been the use of other materials than silicon to create microstructures. LIGA technology can be used to form mechanical
structures in plastic or metal using a combination of X-ray lithography, plating and
moulding. This particular technique has found favour because it allows a larger
variety of high aspect-ratio shapes to be formed [a few microns wide and up to
several hundred microns deep] and although the lithography stage is expensive, the
subsequent electroforming and moulding can replicate low cost components from
one master structure.

Covering the silicon layers of a microstructure with other materials has enabled
a wider variety of sensing structures to be developed. For example, magnetic
materials, piezo-electric materials, electro-optic materials and biosensitive materials

Table 1. Classification of Microsystems and Example

Physical Domain	Input Device	Output Device
Mechanical	Airbag accelerometer	Tilting mirror display
Optical	Micro spectrometer	FED display
Fluidic	Airflow sensor	Ink-jet printer
Thermal	Temperature sensor	Bubble-jet printer
Magnetic	Disk Read/Write head	Magnetic ink printing

Table 2. Applications of Microsystems enabled by Low Power VSLI

Product Application	Technical Features
Cochlear Implant	bio compatible packaging long battery life signal coupling through skin small size
Smart Medical Card	long battery life magnetic or optical data I/O large memory capacity small height components
Portable Navigation System	low cost gyroscope long battery life cellular radio/satellite receiver
Remote environmental monitoring	low power radio link self calibrating systems high stability sensors
Implanted Drug Delivery Systems	reliable fluidic devices low power radio/optical links syringe rechargeable reservoirs

are being used to produce devices such as magnetic disk read-write heads, gas sensors, optical switches and sensors for enzyme-catalysed reactions.

Microsystems can be classified by the particular physical domain in which they operate. Table 1 below shows these classifications and gives examples of microstructures from each.

The combination of smart microsystems with low power VLSI will enable many new products to be developed with large market potential. Table 2 gives examples of such products and lists the technical features which will be required.

2. FABRICATION TECHNOLOGY

It is assumed that the reader is familiar with the basic techniques of integrated circuit manufacture such as photolithography, etching, metal deposition, dielectric deposition and polysilicon deposition. In this section we discuss how these basic techniques have been adapted to produce silicon bulk microstructures, and silicon surface microstructures. We also cover the complementary techniques of LIGA and placing additional layers above integrated circuit structures [so-called ABOVE – IC processes].

2.1 Silicon Bulk Micromachining

This technique involves the removal of most of the bulk of the silicon wafer to leave behind the desired structure on the front surface. It can be contrasted with silicon surface micromachining which leaves the wafer intact but builds up additional silicon layers on the surface. Figure 1 shows a bulk micromachined structure designed to accurately locate optical fibres (photo courtesy of IMC, Sweden).

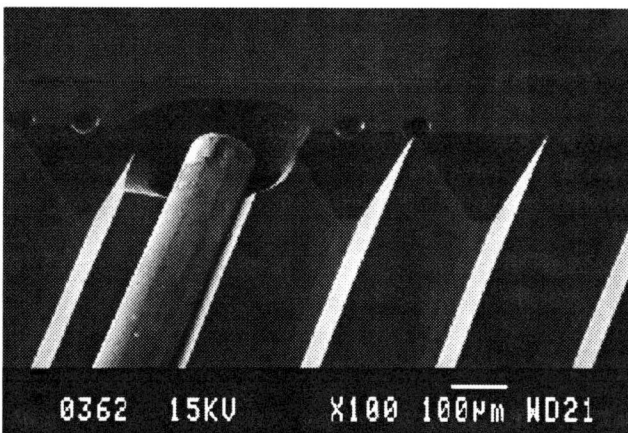

Figure 1. Etched V grooves in silicon

Bulk micromachining takes advantage of the strong anisotropy of some wet etchants with respect to the different silicon crystallographic planes. These etches effectively stop when they reach particular crystal planes. Structures such as v-grooves, beams across cavities, sharp points and diaphragms can be produced.

The particular etches used are either dangerous [poisonous or explosive] or they are incompatible with standard integrated circuit processes. New etches are being developed, such as TMAH [tetra methyl ammonium hydroxide], which do not contain alkali metal ions and are non-toxic. Some more complex structures can be formed by bonding together more than one micromachined wafer [see Section 3.1 Wafer Level Bonding].

Lithography on conventional IC processes has already reached the 0.35 micron level in production. In bulk micromachining, typical structures are around 10x larger for several reasons: special exposure equipment is needed for double sided wafer processing, high depth of focus is needed because of the topography of the surface, and the need for full wafer exposure systems [rather than 5x steppers] to cope with the large device fields.

2.2   Silicon Surface Micromachining

Surface micromachining builds structures on the surface of a silicon wafer. In many cases a sacrificial layer is deposited which is etched away later to create free-standing structures. Typically these layers are 2 or 3 microns thick.

In order to create deep structures new techniques such as reactive ion etching have been developed. Structures in polysilicon, silicon dioxide or other materials as deep as 80 microns have been produced although a more typical depth is 10 to 20 microns.

The main advantages of surface micromachining over bulk micromachining are that the shapes which can be created are not constrained by the crystallographic directions of the underlying silicon and that the processes are generally compatible with CMOS technology so electronics circuits and mechanical structures can be combined on one chip.

The disadvantages are that the properties of some of the deposited layers [e.g. polysilicon] are not as reproducible as the bulk material. This, together with the reduced thickness of structures gives reduced performance in some applications.

Lithography can approach the same limits as conventional integrated circuits, although the resultant structure geometry will depend on the thickness of the layers which are being formed.

Multiple deposited and sacrificial layers have been employed to make complex structures such as micromotors and digital mirror devices.

2.3   LIGA

The LIGA process uses X-ray lithography to produce very high aspect ratio structures in photoresist. The patterns can then be transferred to metal using electroplating and then to plastic or other materials by moulding.

Table 3. Above IC Materials and Applications

Application	Physical Effect	Material
SAW devices	Piezoelectricity	Pb (Zr Ti)O_3
Optical waveguides	Kerr effect	Pb (Mg Nb)O_3
Frequency doublers for lasers	Second harmonic generation	Li Nb O_3
Enzyme immunosensors	Electrical resistance	Carbon
Electronic Nose	Voltage	Polypyrrhole

The main advantages of the process are the very deep structures which can be produced, as much as 0.5 mm but usually not over 100 μm, and the ability to create structures in other materials than silicon e.g. plastics such as polycarbonate or metals such as nickel-cobalt.

The main disadvantage is the high capital cost of the synchrotron which is used to produce the X-ray radiation. However, when LIGA masters are used to produce moulded parts, the tooling cost can often be amortised over a very large production quantity.

2.4   Above IC Processes

In order to broaden the range of physical parameters which can be sensed and increase the functionality of microsystems, many different materials can be deposited and photoengraved above the integrated circuit layers on a wafer.

Some examples are given in Table 3.

3.   PACKAGING TECHNOLOGY

Whereas the conventional IC industry has developed a range of standard packages with associated multiple sources and competitive pricing, microstructures packaging is less mature and presents many technical challenges.

Many microsystems applications create unique demands in terms of the packaging and so application specific solutions are having to be developed. One or more of the following requirements have to be met:
– Isolate sensing elements from non-sensing elements, e.g. fluid access to sensing element but associated electronics hermetically sealed from what might be a harmful environment
– Do not transfer mechanical strain to sensitive mechanical elements
– Provide access for non electrical inputs and outputs, e.g. light, heat, fluids, mechanical coupling.

In addition to these requirements for the finished package, it is important that the packaging process itself does not damage the microstructures or significantly alter their properties.

3.1   Wafer Level Bonding

Bonding of silicon wafers together or to glass layers provides a means of creating sealed cavities or more complex 3D structures.

The sealed cavities can be one means of providing local protection (micropackaging) to sensitive structures or creating reference cavities which are gas filled or in a vacuum. In some cases, the use of micropackaging has allowed the use of a cheap plastic injection moulding process to be applied subsequently without damaging the sensitive microstructure.

The important parameters which define the state of the art are:
– wafer diameter;
– alignment accuracy (wafer to wafer);
– ability to gas fill cavities;
– temperature of the process (minimising damage to microstructures).

3.2   Flip Chip Bonding

Flip chip technology is a means of providing electrical and mechanical connection from chip level devices to an underlying substrate (which may be a larger silicon device, thick or thin film hybrid substrate or flexible PCB). The flip chip process uses a solder bump metallurgy which usually requires changes to the standard metallurgy used on IC's (aluminium). This implies a cost premium which can be justified in applications which have a requirement for small size (thin package) or short electrical connections (no bonding wires). Where these requirements are not so stringent, a multi-chip hybrid assembly can be used.

Advantages of the process are that electrical connections can be formed across the entire area of the device and not just limited to the perimeter as in wire or TAB bonding. Up to several 1000 bonds can be produced in one operation which makes the process useful for connecting arrays of sensors.

Bumping technologies fall into a few simple categories: (1) deposited metal, (2) mechanical attachment and (3) polymer-based adhesives as well as combinations. The original C4 high lead-content solder bumps (melting point >300°C) are being replaced by eutectic solder and adhesives that drop the bonding temperature to a range that is easily tolerated by organic PCB's. C4, however, can still be used on FR4 if eutectic solder paste is applied as the joining material.

3.3   MCM and Hybrid

An MCM [Multi Chip Module] is a way of packaging several unpackaged integrated circuits on a common substrate. The substrate can be PCB [printed Circuit Board] material, thick film printed patterns on alumina, a metal lead frame or a silicon substrate. The different substrate materials are chosen to meet particular performance requirements. PCB is the lowest cost option and silicon substrates are usually the most expensive.

The advantages of MCM are (i) higher density than using separately packaged chips, (ii) ability to combine different chip technologies within one package [e.g. silicon and gallium arsenide], (iii) ability to trade-off chip size to get optimum cost [e.g. 2 smaller chips can cost less than 1 large chip].

The SensoNor accelerometer shown in Figure 1 is an example of a two chip module. MCM's were previously known as hybrid circuits.

3.4   Injection Moulding

Once chips have been mounted on a substrate, it is usually necessary to provide some mechanical protection. This can be in the form of a lid which is glued, soldered or joined to the substrate by a low melting point glass. A lower cost method is to cover the chips in plastic using injection moulding. If the plastic is in direct surface contact with a conventional integrated circuit, there is usually no problem. However, it is often not possible to directly injection mould microsystems because of the sensitive mechanical nature of many of these structures [e.g. accelerometers]. In this case the microsystem is often covered with a cap of glass or silicon using the anodic bonding process described earlier.

4.   DESIGN AND TEST OF MICROSYSTEMS

4.1   Design Methodology

Microsystems technology is relatively immature compared with standard integrated circuit technology and so it is essential to have a well-structured design methodology. Figure 2 shows a typical flow chart used for microsystem development at CSEM in Switzerland.

The development starts off with the system design phase in which the product specifications are partitioned over the different system components. Simulations are performed to verify that the system will meet all specifications. In the next stage, all components are designed in detail according to their specifications. The results of the detailed simulations are cross-checked against the system level simulations. When the components meet the specifications, they are fabricated, tested and assembled to form the first prototype of the system. This prototype is then tested extensively to gain insight in the tolerances of the system to different parameters. When the initial prototype meets all critical specifications, the project continues with the design of the final prototype. Minor modification to the design will be made to assure that this prototype meets all specifications. The experience gained with the fabrication will now also be used to optimise the final prototype so that it can be produced in an industrial way without any further modifications. The product-specific equipment necessary for this future production will also be defined at this stage. The final prototypes are then fabricated, tested, and sent to the customer. They can also undergo the environmental and quality tests specified in the development contract.

The above methodology for a microsystem development is similar to the one for an ASIC, with two notable differences. The first difference is that the microsystem methodology develops an IC, a microstructure, and a package in parallel, with much emphasis on their interactions during the entire development stage. The second difference is that the ASIC development methodology does not distinguish between a first and industrial prototype. The need for this distinction in microsystems stems from the fact that there are no standard test or assembly procedures for microsystems. Therefore, the first prototype is used to optimise the test and assembly procedures for industrial production. The resulting industrial prototype is conceived in such a way that the prototype can be produced in large quantities without the need for redesign in the industrialisation stage.

The system design phase is very important in the microsystem methodology. In this phase, three different issues are addressed. The first issue is the system partitioning that distributes the customer specifications over the different components of the microsystem. The second issue is the choice of technologies for the different components and the verification whether the component specifications can be met with the technologies chosen. The third issue is concerned with the assembly and test of the components and the system. Given the small dimensions of a microsystem, a test and assembly concept must be worked out during the system design.

Throughout the entire methodology, there are checkpoints defined with the precise definition of the information that must be available. The checkpoints are very effective in limiting the risks of the microsystem development, since they require the evaluation of all critical aspects of the microsystem and split the development stage into shorter well-defined parts.

The methodology also helps defining the software tools needed for the microsystem development. The requirements of the software tools for each step of the development stage are based on the kind of information that must be available at the end of the step. This has helped CSEM to choose the software and implement the libraries necessary for each step. The libraries in turn will help shorten the development time and reduce its cost, since they maximise the re-use of available components and knowledge.

4.2   CAD Tools for Microsystems

A number of CAD tools for MEMS exist today, including some low level design tools. While these help in the structural analysis of devices, high level system design tools are needed to provide structured design methods for MEMS.

One approach would be to apply VLSI CAD techniques but these need modifications before they can be used for the automated design of micromachined devices.

There are difficulties. Most VLSI structures are collections of rectangular boxes while MEMS structures are often curvilinear. Orientation and placement of components matter for MEMS where VLSI components can be placed anywhere at any orientation. Visualisation of the third dimension is essential for MEMS design since devices have 3D form and function.

Multi-energy domain simulations are necessary for MEMS and standard cell libraries similar to those of VLSI will need to be developed for systems level design. For example, the mechanical resonance properties of a vibrating structure with a piezo resistive element might be coupled into an electronic circuit.

Now Pasadena based Tanner EDA is currently developing a systems level design package, MEMS-Pro, for use on PC and Unix platforms. The tool software suite has been built on top of Tanner-s existing VLSI CAD tools and includes a layout and schematic editor, spice simulator and MEMS layout, schematic and simulation libraries.

Design of a MEMS device starts with a schematic drawing. A netlist is then extracted from the schematic and used as an input file for running simulations. A MEMS schematic is extracted to separate out coupled electrical and mechanical networks. T Spice is used to simulate the electromechanical performance of MEMS devices. This is a numeric solver of systems of ordinary differential equations that will solve such systems regardless of whether the through and cross variable names are current, voltage, force or position. Simulation of DC, AC and transient performance of electromechanical systems is also possible.

Once simulations meet specifications, physical layout is created with a hierarchical layout editor handling an unlimited number of masks layers, levels and cells and providing for all-angle, 45° and orthogonal drawing. A cross section viewer allows visualisation of a simulated cutaway of a layout in 3D. MEMS designers can program the cross section viewer to simulate the grow/deposit, implant/diffuse and etch steps used in the fabrication process.

Another feature is a user programmable interface allowing automatic parameterised generation of layout.

Tanner is now working on 3D solid model generation, and viewing from the layout editor with 3D analysis consisting of simultaneous electrical, mechanical, magnetic and thermal analysis.

Workers at CMP in France have extended the capabilities of CADENCE OPUS to microsystem technology. The user is able to generate layout of a microsystem including electronic and non-electronic parts in a commonly used format [GDS2, CIF] run an extended DRC which submits the layout to design rules checks in process aspects and electrical aspects and, finally, extract parameters from the layout level to the netlist level. The extended extractor will recognise electronic components as well as microstructures, such as bridges, cantilevers and membranes.

A netlist is generated and the simulation can be executed by the means of parameterized behavioral models of these microstructures.

4.3   Testing Issues

Testing and packaging of microsystems contribute a considerable proportion of the total cost. Estimates vary between 40% and 75% of the total cost.

In many cases, the full function of a microsystem can only be tested after it has been completely assembled [e.g. pressure sensors, micro-optical systems or

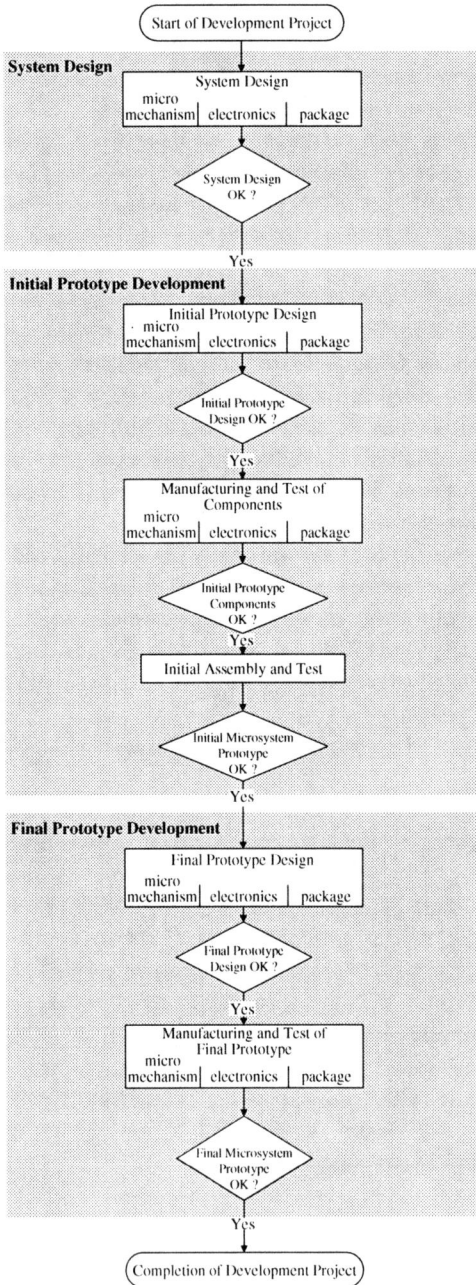

Figure 2. A typical flow chart used for microsystem development at CSEM

micro-fluidic systems]. This is in contrast to conventional integrated circuits where an almost complete functional test can be carried out at wafer probe level before packaging.

To reduce the cost and increase final product yield, it is important that some intermediate tests at the wafer level can be developed which will give a good indication of functional yield. Some examples which have been developed for accelerometers indicate how this might be done: tilting probe tables allow the earth's gravity to deflect proof masses; electrostatic forces can be used to deform beams and then measurements of deformation using capacitive coupling can be made.

New test structures will need to be developed whose parameters track closely the required functions for valves, pumps, comb actuators, tilting structures, microphones, gyroscopes, etc.

Chemical and biological microsystems present special problems because the sensors usually require some form of external protection or activation.

Magnetic sensors relying on magnetic-resistance, hall effect or inductive coupling, are usually only tested on electrical parameters until after packaging because of the difficulties of producing the required magnetic fields at the probe level.

Optical components require assembly with very tight mechanical tolerancing. Even components which have been 100% tested before assembly do not guarantee system functionality without active alignment. Some laser devices cannot be tested until they have been cleaved from the wafer because the cleaved face forms part of the optical cavity.

5. APPLICATIONS FOR MICROSYSTEMS

Microsystems products are already finding application in many areas and will eventually penetrate most of the markets which are using conventional integrated circuits as well as stimulating some new ones. A recent study by Technology For Industry Limited has identified applications in the following areas; automotive, medical displays, sensing and industrial process control, communications, computer memory, printers, instrumentation and environmental monitoring.

5.1 Automotive Applications

The automotive market is already a major user of microsystems [$20 million for accelerometers, $400 million for pressure sensors] and is expected to become increasingly important because of the variety of new products in development and the steady growth of existing applications. Accelerometers are currently used to trigger airbags but will increasingly be used in skid control and active suspension systems. Pressure sensors are used for manifold air pressure measurement and will move into oil pressure, tyre pressure and other applications.

New optical sensors will be used for rain sensors, fingerprint keys, smart headlights and climate control [sunlight]. Radar sensors are planned for proximity warning devices. Rotation sensors are already in ABS systems and will be incorporated steering wheel measurement.

Yaw sensors [angular rate] will be used for skid control, navigation systems and active suspension.

Cost is extremely important in the automotive market and microsystems may need to combine two or three sensors in one device for less than $5 in order to replace conventional sensors.

5.2   Medical Applications

Existing high volume products include blood pressure transducers and hearing aids. High value but low volume markets include pressure sensors and ultrasonic probes in catheters, cochlear implants, accelerometers in pacemakers and DNA replication systems.

The level of research and development in this area is high but the requirements for extensive clinical testing increase the time to market for many products. It is anticipated that drug dispensing and monitoring systems, inner-ear hearing aids, smart catheters, patient monitoring systems, pain relief by nerve stimulation and DNA analysis systems could all become important applications for microsystems.

5.3   Instrumentation and Process Controls

Industrial applications are extremely diverse reflecting the nature of the industry itself. Although certain generic applications such as pressure sensing are likely to be used in a variety of contexts, there will also be numerous uses which are specific to very narrow sectors of the market. In order to derive economies of scale sufficient to develop cost effective solutions, standardised or generic products are being tailored to produce niche products, e.g. Gas/Chemical sensors, Temperature, Pressure, and Flow sensors. Although the industrial market is forecast to become very large, its growth is not expected to be as great as the Automotive industry. Frost & Sullivan projected the Industrial and Process control market size to be $314 million by 1998. The Automotive market is already around $1000M and is projected to reach $3000 by the year 2000 (R Grace Associates). Sectors that have a record of IT update are likely to be candidates for MST, e.g. Oil industry – Pressure & Flow, Chemical industry Pressure & Flow and Chemical sensors. Power generation – numerous sensing requirements, Mining industry – flammable gas, temperature and pressure sensors.

Companies specialising in horizontal applications (application specific system integrators) will sell across a number of industry types. Thus companies selling chemical sensors will be able to sell to oil, chemical, environmental users or in the case of Motorola to the consumer and automotive industries.

Conversely, there will be another tier of companies operating at an even higher level of integration, to provide complete industrial process systems. The level of industry infrastructure, co-operation and understanding are likely to contribute to a slow roll-out of MST products. However a number of profitable niches will develop quickly and could be addressed by SME's.

5.4   Domotics

The consumer market is potentially a significant market for microsystems, and is typified by high volume, low value mass markets. Examples include:
- Pressure sensors used in White goods, e.g. coffee machines, vacuum cleaners, and washing machines;
- Temperature sensors in microwave ovens;
- Various motion and other sensors for electronic products e.g. cameras, security devices, virtual reality games etc.

Within building control for both domestic and industrial environments various sensors are required in Heating Ventilation and Air Conditioning e.g. Temperature, humidity, carbon dioxide, oxygen & flow.

Alarm sensors are also needed to warn of the presence of methane, carbon monoxide and combustion products.

5.5   Computer Peripherals

Microsystems have already had a major impact on the computer printer industry with bubble-jet based products dominating the market and continuing to take market share away from laser printers. It is believed that the market for disposable ink-jet heads is around $2 billion per annum.

Magnetic Disk Read/write heads are expected to become a huge market for microsystems as the new generation heads incorporate combined sensing and signal processing devices. This could be a $4 billion plus market for microsystems.

5.6   Displays

Displays represent a $20 billion plus market with TV's, PC monitors and portable computer flat-panel displays being the highest volume sectors. New microsystems based devices such as field emission devices [FED's], ferroelectric displays, interference displays, grating light valves and digital mirrors are all attempting to enter the display market and replace conventional displays such as CRT and LCD.

6.   FACTORS LIMITING EXPLOITATION

Despite the tremendous potential of microsystem technology, the take up has not been as rapid as might have been expected. In this section we will look at some of the factors which have acted as barriers to the use of the technology.

6.1   Lack of Clear Interface between Microsystems Designers and Manufacturers

The specification and design interface between user and manufacturer is complicated by the multi-domain nature of microsystems [electrical, optical, magnetic, mechanical etc.]; the lack of simple design rules or cell libraries; and the special packaging and test requirements which may be required. This is very much an issue of technological maturity, and is compounded by the lack of CAD tools.

6.2   Need for Multi-Disciplinary Skills at the User Site

In many cases, particularly where microsystems enable new products to be conceived, the design involves multi-disciplinary efforts. This is not just because of the multi-domain nature of microsystems but because the microsystem may integrate devices of functions which were previously defined an implemented by separate groups of engineers. This multi-disciplinary competency may be especially difficult to achieve for SME's.

6.3   Complexity of the Supply Situation

Users are faced with a bewildering array of technology offers from a large number of suppliers. In most cases the suppliers are not able to offer all the technologies and offers from different suppliers are not in a form which can easily be compared. Technologies are described in terms of technical features rather than the economic or other benefits which they potentially offer the user.

6.4   Lack of Access to High Volume Production

Most microsystems technologies are on offer from research institutes which are not able to provide high volume capability which is needed to achieve most cost effective production economics and help mature the technology by learning curve experience. Those production lines which are able to achieve high volume are usually within vertically integrated companies and are not open for external customers [e.g. automotive or aerospace companies with captive facilities].

6.5   Perceived High Entry Cost

The design and manufacturing tolling costs, even for established microsystem technologies, are high, especially for SME's. Where new processes are required, the development costs are too high even for many larger companies.

6.6   Perceived High Risk Level

Except for the airbag application, very few successful microsystems developments have been widely publicised and hence many companies will adopt a wait-and-see attitude until the benefits of the new technology are demonstrated by another company.

7.   FACTORS ENCOURAGING EXPLOITATION

In response to the problems identified in Section 6 several National or International initiatives have been started to encourage companies to use microsystems technology. These include the EUROPRACTICE and FUSE projects in Europe, the MIB project in the UK, funding from DARPA in USA, MITI in Japan and BMBF in Germany.

7.1   EUROPRACTICE

EUROPRACTICE is a 4-year project with a 34 million ECU budget [of which just over 50% is for microsystems] funded by the European Commission. The project started in late 1995 and its objective is to help create an infrastructure to support companies who wish to use microsystems technology. This is important for companies who need access to design skills and manufacturing facilities which they do not have in-house.

EUROPRACTICE has established Manufacturing Clusters which offer established microsystems production processes which can handle medium or high volume. This reduces the risk of companies developing a prototype on a research and development line which is not compatible with high volume production. In addition, the project is stimulating the development of fab-less design houses which can handle the interface to the manufacturing foundries.

7.2   FUSE

FUSE is another European Commission supported project which provides up to 150K ECU in project funding to companies who already have an existing product or market presence, but want to become more competitive by using more advanced electronics technology. In return for the project finance, the companies must agree to share their experiences and encourage others to follow their example.

8.   BUSINESS STRATEGIES FOR MICROSYSTEMS COMPANIES

In most sectors of business, there are three competitive strategies which can be adopted: to be a cost leader, to select a performance niche or to provide high added value and service. The situation for companies wishing to get involved in microsystems is no different. The main problem is to get access to the right information to enable the company to choose the most appropriate strategy. This is particularly the case in microsystems where the technology, the players involved and the markets are developing so rapidly.

8.1   Cost Leadership

Once a market application matures to the point where a standard specification has evolved [this might be the result of international standards committees work or de-facto standards emerging] then a commodity market can be created. Several manufacturers produce devices to meet the same specification and the main distinguishing factor in the eyes of the customer will be price. In the microsystems area this had already happened with many kinds of pressure sensors and some kinds of air-bag acceleration sensors. Motorola are set to become major players in both these markets and have already made some very aggressive projections about the low prices they will be offering. Many smaller companies in these markets will find it increasingly difficult to survive when faced with the massive investments that Motorola is making, both in manufacturing capacity and advertising budget.

8.2   Niche Selection

One response to the falling prices in the commodity markets is to move into niches in the market where additional performance is required. This is the response of many of the pressure sensor manufacturers. For example, Silicon Microstructures Inc. have recently introduced an extremely sensitive device with 0.15 psi full scale. In the accelerometer market, Analog Devices Inc. have moved from single axis devices to combining dual axis devices on one chip which gives more value to the customer.

In the display devices market, many of the new technologies are not being used to compete directly with the CRT or LCD [at least, not at first] but are targeted at a segment of the market such as high-brightness displays for aircraft, or small high-resolution displays for virtual reality goggles.

In contrast, Pixtech which has developed the FED technology is attempting to move quickly into high volume markets by licencing very large production companies such as Futaba and Motorola.

8.3   High Added Value

Several microsystems devices are very expensive to develop but provide very significant competitive benefit to the systems within which they are incorporated. For example, the DMD device [digital mirror display] is only a small part of a system used to project high brightness images in the home or business environment. Texas Instruments, which developed the DMD device, is attempting to recoup its investment by selling complete subsystems [light source, optics, power supply etc. plus DMD device].

9.   CONCLUSIONS

In order to develop a successful business strategy, a company will need access to the right information about markets, technologies and investment finance.

Market information is crucial to ensure that products are developed which fulfil a real user requirement and the most effective channel to market is selected.

Technology information enables a company to choose the technology which will give the best competitive edge whilst meeting the required functionality. It is also important to choose the most appropriate level of integration to match the expected volume requirement and cost targets.

Finance levels must be appropriate both for development and manufacturing. Without adequate investment in development resources, market windows can be missed or a small company starting early can be overtaken by a large player coming in later with superior resources. Manufacturing investment in capital equipment may be essential to achieve high volume production at competitive prices.

CHAPTER 3

CORE ARCHITECTURES FOR DIGITAL MEDIA AND THE ASSOCIATED COMPILATION TECHNIQUES

JOCHEN A.G. JESS

Eindhoven University of Technology, P.O. Box 513, 5600 MB Eindhoven, The Netherlands,
Phone: +31 40 2118041, e-mail: J.A.G.Jess@planet.nl

Abstract: The new generation of multimedia systems will be fully digital. This includes real time digital TV transmission via cable, satellite and terrestrial channels as well as digital audio broadcasting. A number of standards have been developed such as those of the "Moving Picture Experts Group" (MPEG). Those are defined for the source coding of video signals. Various channel coding standards for audio and video transmission based on spread spectrum technology have been established by the "European Television Standards Institute" (ETSI). While the video receivers will come to the market in the form of set top boxes feeding standard PAL-, SECAM- or NTSC-receivers with the appropriate analog signals, a new generation of mobile TV- and radio receivers on the basis of mobile phones and PDA's is also conceived. One of the central questions regarding the implementation of those products is the balance between flexibility and speed performance of the various hardware-platforms. The competition is between standard processors like the new generations of the Pentium (as "superscalar" representatives), possibly enhanced with add-ons like MMX™, or, alternatively, the TRIMEDIA, (a VLIW representative). Somewhere on the scale resided the "Digital Signal Processors" like the TSM 320. But also special purpose architectures like the PROPHID architecture have been developed for applications like multi-channel real time video or real time three dimensional graphics for "virtual reality". Late in the nineties the PROPHID architecture has been cast into an experimental prototype called "CPA" (Coprocessor Array). All those architectures need special software techniques to actually exploit their potential in terms of operation speed. The paper reviews some of the relevant features of the signal streams and the sort of processes that have to be executed on them. Then it discusses some hardware architectures that compete as media-processors. Eventually the question of software design for such architectures is addressed culminating in the description of some recently discovered scheduling techniques to be used in compilers/code generators for those processors

Keywords: Multimedia systems, embedded systems, VLSI-technology, core-processors, intellectual property, compilers

27

R. Reis et al. (eds.), Design of Systems on a Chip, 27–63.
© 2006 *Springer.*

1.   INTRODUCTION

In 1994 the "Semiconductor Industry Association" (SIA) published a document
called the "SIA Roadmap" defining the key objectives for the monolithic semiconductor fabrication technology all the way to the year 2010 [ITRS]. According to
this "Roadmap" in 2010 the standard DRAM chips would store 64 Gbit. A microprocessor was predicted to contain 90 Million transistors (excluding the on-chip
cache!). A chip is said to possess up to 4000 pins and would be approximately
four by four centimeters in size. It would run on a 0.9-Volt power supply possibly
with an 1100 MHz clock. Currently the roadmapping activity has been extended to
world scale. The technology data are significantly refined and continuously updated.
Industry as whole developed a tendency to outperform the roadmap, which in turn
caused the roadmap planners to adjust their targets to more aggressive numbers.
Recently the numbers show a tendency to consolidate. The 2003-roadmap indicated
for 2014 that standard DRAM technology will offer chips holding 48 Gbit on an
area of $268 \, mm^2$.

   With its initiative the SIA triggered other industries to engage into making plans
for the future. It is safe to assume that the plans of the microprocessor vendors have
been fixed for the period in question. The same goes for the other semiconductor
vendors. Other industries follow suit such as for instance the Consumer Electronics
industry and the computer manufacturers. Both of the latter have ambitions regarding the media industry. The computer industry wants the "PC" to become a device
connecting its user to the rest of the world integrating data, sound and vision into
one powerful communication channel. The Consumer Electronics industry wants
the TV-set to become similarly powerful. It can be predicted that eventually the
consumer/user may demand some standardization of the "look and feel" of the
user interfaces in order to master the complexity of the devices. But before those
issues are settled an enormous variety of products will enter the market. This
market will be extremely competitive because of the inevitable "shake-out" to be
expected.

   In the meantime systems- and chip-designers worry about how to master the
complexity of the design process for the new media systems in view of the potential
of the semiconductor industry. The new chips will integrate a full system on one
die including a multitude of processors of all kinds. Also included will be a large
variety of memory devices, programmable logic and programmable interconnect,
all kinds of channel decoders and modems and even radio frequency front ends
in the Gigahertz range. The designers will have to deliver the respective designs
within extremely short time frames, typically within few months. In turn this implies
very efficient design environments. Yet the growth in productivity of the design
environments has been lagging behind already over a long period of time. The
growth does not exceed one third of the growth of the potential of integration
achieved by the semiconductor manufacturers (see [HUR96]). This means that if
nothing changes in design technology the lead times grow beyond any acceptable
limit. Eventually there might be not enough products delivering revenues from the
investment into the new semiconductor technology.

The answer is sought by referring to three concepts, namely:
- standardization;
- reuse;
- programmability.

The basic idea is that only a few generic processor architectures, so-called "processor cores" will be designed covering a wide variety of applications. They are supposed to be held in stock by specialized vendors as "intellectual property" (IP). This IP can be purchased, say, via Internet together with software that performs the mapping onto the customer's die in compliance with the customer's semiconductor manufacturing technology. Also with the IP comes a development system that helps to develop code to be downloaded into the Silicon structure.

It may be fairly obvious to the reader that we have yet to achieve a profitable implementation of this concept in the industry despite the fact that companies like ARM Systems have started a commercial operation along these lines. (The acronym "ARM" stands for "Advanced RISC Machines".) But CAD-vendors like CADENCE or SYNOPSYS adjust their marketing strategies according to this concept increasingly.

Technically there is a lot still to be done. First of all there is a wide open question concerning the most appropriate cover of architectures for the functions to be handled by the cores. The competition is just opened and there is no consensus at all about the answer. This question will take years to settle. The emerging standards will be the de facto standards imposed by the winners of the competition. Standards will also come under pressure because the development of applications is moving fast (see for instance the move towards MPEG4 video operation). This is inevitable, as it is part of the dynamics of the market and thus an element of the driving force of the industry.

Secondly the concept relies on a certain degree of standardization of the manufacturing technology. If the fabrication processes and the associated design rules develop too much spread it becomes too difficult to prepare valid footprints of the cores on an economical scale. It is a fact that the manufacturers are currently subject to numerous reorganizations including a lot of mergers and license agreements (see for instance the earlier cooperation between DEC and Intel or the one between HP and Intel, but also the cooperation between Philips and SGS-Thomson and some Japanese companies). But it is still not entirely clear that this will yield standards sufficiently stable and uniform to alleviate the making of the very complex layouts for the cores matching a variety of fabrication processes. There is also the problem of the very fast evolution of the fabrication technology (see the Technology Roadmap!). The question will remain whether the cores will "scale down" efficiently due to the more pronounced influence of physical phenomena, as the feature sizes of the Silicon structures scale down. The issue of manufacturability enters the discussion ever more frequently.

The limits to standardization will therefore impair the reuse of the cores. How the market forces arrange the economic situation around the cores is difficult to predict. Programmability, however, presents itself in a different fashion. The computer industry can look back on an overwhelmingly successful history while building its products on programmable platforms. Contrarily the Consumer Electronics industry has a strong tradition in making complex products like TV-sets, video cameras and DVD recorders from "Application Specific Integrated Circuits" (ASIC's). Only recently numerous standard microprocessors and Digital Signal Processors (DSP's) have entered the TV-set. Currently the software development cost for a TV-set dominates the hardware development cost, even for analog TV technology. The advent of digital TV will enhance this tendency even further. Current set top boxes are composed from a set of standard cores combined with ASIC's (giving them the performance of a supercomputer) and are highly programmable to serve a sequence of product generations. The next generations will integrate more and more functions on one chip.

In the sections below this paper will extend on the programmability issue. First some features of the data formats in the media will be reviewed. Next the discussion focuses on features of processor architectures and the consequences of those features for the development of the associated (embedded) software. This discussion culminates in a set of requirements concerning the development systems for those processors and their impact on the compilers. Compilers will have to serve more stringent performance requirements but in turn are granted more time to perform the necessary optimizations. Consequently the problems of binding, allocation and scheduling have to be reconsidered in the light of the changed conditions. In particular new techniques for scheduling are presented and their performance is illustrated.

It is expected that those techniques will remain significant no matter which way the cores develop. The problems of optimization are fundamental yet not addressed in this way in the past, as the applications were not in the scope of processor design in earlier phases of history.

2.   STREAMS OF MEDIA DATA AND THEIR PROCESSING

In the history of computer engineering the use of the computer is pictured such that a user submits a program that is subsequently executed by the computer. This basic pattern has essentially been preserved through the entire evolution of the computer architecture regardless whether we consider the multitasking systems of the mainframe époque, the first stand alone personal computers, the workstation era or the distributed computer networks of today. The basic architecture experienced by the user is still the von Neumann machine although in fact it may assume any degree of abstraction: it is a virtual machine hidden behind a complex user-interface. The level of acceptance through the user determines the performance requirements. Being aware that he has submitted a computationally demanding job he is ready to wait an adequate amount of time as long as he thinks that eventually the job will be successfully completed.

Consequently the pressure on the developers of computer hardware and software was more on reliability, as the semiconductor technology gave plenty of speed – more than necessary for a single user. The CISC (Complex Instruction Set Computer) and the RISC (Reduced Instruction Set Computers) architecture concepts provided speed performance sufficient to service hundreds of users simultaneously from a single hardware platform. When at the end of the seventies the "software catastrophe" was threatening this was essentially a reliability issue induced by the regular appearance of software errors.

Real time video and sound applications on the personal computers brought the time performance issue back onto the agenda of the computer manufacturers. In addition to the real time requirements imposed by on line video operation the portability of much of the equipment opened two other dimensions of performance: weight and power consumption.

Perhaps a little to the surprise of the computer manufacturers the Consumer Electronics industry started developing the receivers for digital TV-transmission all by its own. The first new product in this area was the so-called set top box. This device converts the digital video signal streams into standard analog signals accepted by the PAL-, SECAM- or NTSC-TV sets now standard in the homes. The performance of the set top boxes is enormous by all standards. With the current products it reaches tenths of Giga-operations per second (GOPS). Even in today's semiconductor technologies the chips are small and their power performance is some hundred times better than that of a standard microprocessor. (It must be admitted that most processors are not even able to produce real time digital video of the visual quality that a set top box equipped with special hardware chips reaches easily.)

The differences in performance are of course not so difficult to explain. The chips in the set top box are tuned to certain specialized jobs very efficiently. For instance the so-called MELZONIC processor serves to convert 50 frames per second interlaced TV into 100 frames per second progressive operation within a modern high end (100 Hz) analog TV-set currently on the market [LIP96]. The MELZONIC processor computes intermediate pixels by an interpolation technique resembling the computation of motion vectors in the MPEG2 encoding procedure. The one million transistors on this chip perform this task at a level of 9 Giga-operations per second – 300 operations in parallel at every clock cycle. Contrarily the TRIMEDIA TM1 processor running at 100 MHz is able to perform a simplified version of the MELZONIC algorithm. He can do that in real time at a rate of 0.1 Giga-operations per second. The TRIMEDIA processor is then loaded for 95% of its capacity, it is three times as big in terms of area and uses about twenty times as much power [SLA96]. But it has the advantage of being able to serve other jobs by changing his programs, a feature not available with the MELZONIC.

The reason why chips like the MELZONIC can be so efficient is closely related to the data standards fixed for on line video operation. Essentially all data representing picture information are put into packages that are transmitted bit-serially. Any package contains a header and a payload. The header tells all necessary details for the processors to deal with the payload. In the MPEG2 standard the header- and payload-portions of each package are standardized ([ISO93], [ISO94], [ISO94a],

[ISO95], see also [LEG91]). Packages may be nested according to a carefully designed hierarchy. In MPEG2 the highest level of the hierarchy is the "group of pictures". This is a sequence of pictures delimited by context- (or scenario-) switches such as cuts in a movie. A packet representing a group of pictures consists of "pictures". A picture in turn consists of "macro-blocks" of 16 by 16 pixels each, which cover the whole picture like a chessboard. Macro-blocks are again split into four 8 by 8-pixel "blocks". Also for certain operations macro-blocks may be lined up to establish "slices".

The MPEG2 (source-) encoding process applies sub-sampling techniques amounting to an acceptable loss of information. Usually the chrominance information is sub-sampled, as the human eye is less sensitive to the loss of color information compared to a comparable loss of intensity information. More substantial is the use of both the spatial correlation within one picture and the temporal correlation within a group of pictures to reduce the amount of bits actually to be transmitted. The spatial correlation is exploited by computing a two-dimensional spatial Fourier transform on the luminance and the chrominance values of each block in a picture. The result is a two-dimensional spatial line spectrum defined on the matrix coordinates of a block. The (0,0) entry of the luminance spectrum gives the average gray value of all the pixels in the block. A high value for the (8,8) coordinate of the luminance spectrum indicates a two-dimensional sinusoidal fluctuation pattern between black and white changing from cell to cell (again comparable to a chessboard pattern). The latter is seldom present in a block. Rather the high frequency spectral lines seem to converge to zero fairly fast. So the encoder lines up the samples in a zigzag fashion starting at the upper left coordinates of the block. The appearance of many zero values in the spatial spectrum is taken advantage of by transmitting so-called (value, run)-pairs where "run" indicates the number of zero values to follow the nonzero valued sample "value". This procedure is called "run length encoding". The technique also normalizes the "value" samples relative to the average gray value (in the case of the luminance encoding) and then applies "Huffman"-encoding to optimally use the available wordlength. This procedure is known under the name "variable length encoding".

The temporal correlation is exploited in the following way: at certain points in time an integral picture is spatially encoded and transmitted (this is the so-called I-picture). The next series of pictures is computed by comparing the pictures to be transmitted next with the I-picture. Just the differences are actually transmitted, this way yielding the B- and P-pictures. P-pictures are computed from preceding I-pictures, while B-pictures are in turn computed by interpolating between I- and P-pictures. Those pictures may largely be composed by blocks resembling those of the original I-picture (or the corresponding I- and P-pictures), which are laterally displaced. Therefore the transmitter, in addition to computing the differences, also computes a vector approximating the displacement that the block undergoes in time. This vector is called the "motion compensation vector".

The computations described so far all refer to the source encoding process, which compresses the bit-stream to be transmitted. In order to ensure reliable transmission over the various channels more encoding is necessary which this time amounts to the addition of redundancy. Both Viterbi and Reed-Solomon encoding are applied. Also the technique of interleaving is applied. This technique implies the partition of the packets of a bit stream into equally sized blocks. These blocks are then reshuffled by a system of delay lines. The idea behind this technique is that a burst contaminating a long sequence of consecutive bits is pulled apart and distributed over various portions of the packet. This in turn reduces the probability for the Viterbi and Reed-Solomon decoders to fail because of sequences of contaminated bits becoming too long. Figure 1 displays the structure of an MPEG2 transmitter for terrestrial transmission.

The encoding is fairly complicated because with terrestrial transmission every tenth bit is lost. On the other hand the European transmission standards require an error probability of less than 10^{-11}. This amounts to about one hour of uncontaminated picture sequences. The transmission technique also implies the application of the principle of "Orthogonal Frequency Division Multiplex" (OFDM), sometimes referred to as "Spread Spectrum Technique". With this technique the digital video signals associated with each individual transmission channel are distributed over the total available frequency range (rather than being concentrated in some small frequency band, as is the case in today's analog transmission systems). This method curbs the impact of "bursts" of contaminating signals concentrated in a coherent portion of the frequency domain.

Figure 1 gives an impression of the integral channel encoding procedure the way it may be implemented in the future digital TV–systems. The systems for satellite and cable transmissions are actually conceptualized as subsystems of the schematic displayed in Figure 1.

In summarizing it can be concluded that media processes deal with streams of data in standardized formats. On the higher levels of the syntactic hierarchy of the video signal stream the lengths of the data packets are fixed by standard values. For instance the PAL standard requires 25 fields per second, 50 frames per second, 720 pixels per row and 576 pixels per column. So in the programs capturing the behavior of the encoding process on this hierarchic level loop boundaries tend to be fixed and there are not too many conditionals depending on run time data. However, as the source encoding process proceeds more and more dynamic decisions take place involving loop boundaries and conditionals depending on run time data. Special purpose architectures can use these circumstances in their advantage both with respect to the interconnect- and the memory structure. The above statements also hold for digital audio processing be it that the demands on speed performance are less pronounced. The problem with handling audio data streams is more related to the high degree of multiplexing involved in the processing. Even processors with moderate speed performance can handle an audio stream in real time. So either many audio streams are processed simultaneously or audio streams are interleaved with video streams very much in the style TELETEXT is taken care of in the analog TV-sets in Europe.

MPEG-2 stream

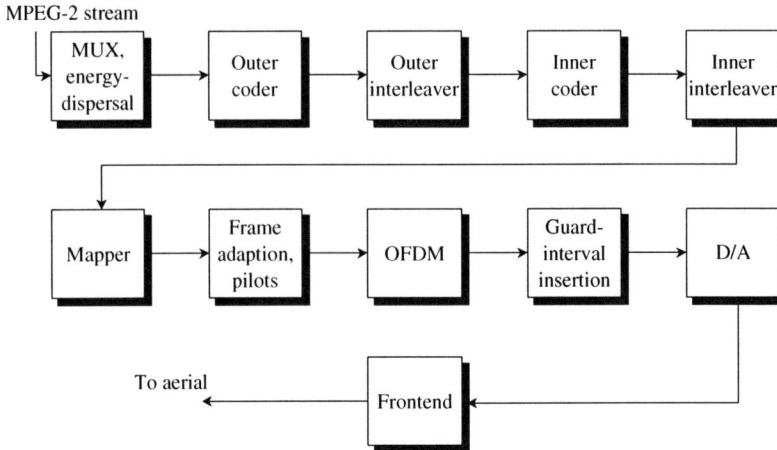

Figure 1. Schematic of an MPEG2 encoder for terrestrial transmission

The computations executed on the streams involve filtering (which is "multi-ply with constants and accumulate"), Fourier-transforms, computing averages and maximum values, truncation and rounding, sorting and matrix transposition, pattern matching and the like. This fact will have consequences with respect to the choices made when evaluating the optimization techniques to be described later.

3.   A REVIEW OF ARCHITECTURES

In principle three different approaches can be distinguished. The first approach starts from the concept of a general-purpose microprocessor such as for instance the Pentium processor. The second approach takes the concept of a DSP as a point of entry. Finally special purpose processors especially made for media algorithms represent the third class of architectures.

3.1   The MMX™ Extension to the Pentium Processor

The objective of the MMX™ extension is to provide improved operation speed for computations typical for media applications without change of the essential architectural features of the Pentium processor [WEI96]. Consequently the extension had to fit into the existing Intel architecture. This implied no additional machine states and no new events. In the original Pentium architecture there was plenty of unused space for operations code. In order to account for the typical data types in media applications the 80/64 bit floating point registers formed a good basis. Those data types were chosen to be the following:
– Packed bytes, mainly for graphics- and video-applications;
– Packed words, mainly for audio- and communication applications;

- Packed doublewords, for general purpose use;
- Quadword of 64 bits for bit-wise operation and data alignment.

Altogether 57 new instructions have been created. They are essentially of the "Single Instruction Multiple Data"-type (SIMD) or otherwise of the type "Fixed Point Integer". They map directly into eight floating-point registers by direct access. Examples of instructions are:

- "saturating add" to compute four sums of two packed words each saturating on maximum overflow;
- "packed multiply and add" multiplying four pairs of packed words and then adding two products pair-wise to form two sums (which actually allows for complex multiplication in one single instruction);
- "parallel compare" of four pairs of packed words by a single instruction yielding a mask of four packed words;
- "Data conversion" which takes two packed doublewords and maps them into four packed words.

The technology is said to achieve a speed-up of a factor from 1.5 to 5 as compared to the Pentium architecture not supported by MMX™.

3.2   The TRIMEDIA Processor

The heart of the TRIMEDIA processor is a synchronous CPU with parallel pipelines allowing for the execution of five instructions simultaneously [SLA96]. The pipelines are controlled by means of one controller with a large number of control lines – therefore the CPU falls into the category of VLIW processors. The CPU contains 32 Kbytes instruction cache and 16 Kbytes dual port data-cache. A 32-bit bus connects the CPU with the external Random Access Memory. Additionally the processor contains a large number of interfaces for communication with media components". In the original version (the T1) there is for instance a CCIR601/656 YUV 4:2:2 video input with 19 Mpix/s and a video output with 40 Mpix/s. Audio interfacing is obtained by a stereo digital audio (I2S DC-100kHz) input- and an -output channel (2/4/6/8 channels). Also there is an I2C bus for cameras, a synchronous serial interface (V.34 modem or ISDN front end) and a PCI bus interface (32 bits, 33 MHz). A "VLD Coprocessor" takes care of the MPEG 1&2 Huffman decoding at slice level. An "Image Coprocessor" provides the down- and up scaling of YUV-versus RGB-format. This happens at a rate of 50 Mpix/s. The TRIMEDIA processor can also be equipped with a special media instruction set.

The processor performs the MPEG2 decoding process of the "Digital Versatile Disk"-player (DVD-player) in real time with a clock period of 100 MHz. This process occupies 64% of the processor capacity. Measured by means of the so-called "DVD-batman" bitstream it achieves on the average 3.95 useful RISC operations per VLIW instruction. This implies an equivalent of 2.9 operations per clock-cycle after correcting for cache misses and conflict stalls. The memory bus is loaded for 18% of its capacity. These are averages. Therefore under top load conditions degradation of the picture quality may occur. With the batman test series (500 frames) 44 of these frames present a load higher than 85% of the CPU capacity. This means that in fact for the DVD operation eventually an upgraded processor of the TRIMEDIA type with higher performance must be utilized.

For a 44.1 Ksample/s audio stream (16 bit MPEG L2 stereo) the CPU is loaded with 4.6% of its capacity while the load of the memory bus is 0.95% of its capacity. In that case the average number of operations per cycle is 2.4.

To obtain an idea of the performance of the TRIMEDIA processor relative to a standard microprocessor the following benchmark, a three-dimensional triangle mapping process, is considered. The three corner points of a three-dimensional image of some triangle establish the input of the process. Three spatial coordinates, four color values and the luminance value describe each point. This amounts to a total of 30 floating-point values. The output is constituted by 44 fixed-point values per triangle, which are off loaded to an off-chip raster engine. Including all loads and stores the process comprises 410 operations. The code for the TRIMEDIA processor without multimedia options can be executed in 210 cycles. With multimedia options the code can be scheduled in 96 cycles. For the Pentium P6-200 this process takes 2440 cycles. It should be noted that the triangle mapping process cannot take advantage of the MMX™ instructions.

3.3   The MISTRAL-, PHIDEO- and PROPHID-Architectures

The MISTRAL-, PHIDEO- and PROPHID-architectures are all especially designed for media applications. While MISTRAL is oriented towards audio- and control-processes the latter two are designed for the implementation of high-speed signal processing tasks such as real time video encoding and decoding.

All three architectures are the outcome of earlier research on Silicon compilation (see, among others, the CATHEDRAL Silicon compilers from IMEC) [MAN90].

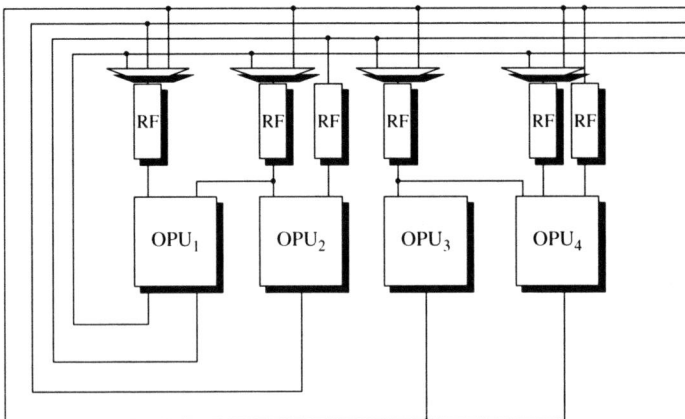

Figure 2. MISTRAL: basic architecture

The concept of a Silicon compiler was established in the early 80s. In principle a Silicon compiler took a process definition as input and created a chip-architecture optimized for the process in question. The CATHERDAL concept started from an architecture template called the "target architecture". The MISTRAL-, PHIDEO- and PROPHID-architectures all are based on data flow templates. As an example in Figure 2 the basic template of the MISTRAL processor is shown [STR95]. It consists of a set of operational units (OPU's) operating in parallel. By way of a structure implemented by a set of buses and an optimized pattern of multiplexers and register files the principal loop is established. In the case of MISTRAL the input process generally is a set of parallel (possibly communicating) processes. The idea is that the individual processes share the components of the architecture. Eventually all components of the architecture contribute to the execution of any of the processes.

The procedure creating the feedback structure attempts to minimize the number of OPU's, buses, multiplexers and register files while simultaneously satisfying the throughput demands of all the individual processes. In Figure 3 the result of such an optimization is shown. In this structure we learn about the type of components that may qualify for an OPU. It can be realized that RAM and ROM are special cases of an OPU in the same way as multipliers, adders or controllers can appear as such. The OPU's are in general highly parametrizable in order to optimize the performance of the structure both in terms of time performance and chip-area. The example of Figure 3 for instance exhibits the case where the multiplications obviously involve pairs of signal samples and filter coefficients exclusively. The filter coefficients are stored in the ROM and routed directly to one of the inputs of the multiplier. The MISTRAL optimizer further decides that one register satisfies to buffer the coefficients. Simultaneously the optimizer plans a stack of four registers to store signal values at the other input of the multiplier where the samples either come directly from the input or are read from the RAM. It is also obvious that the products are supposed to be accumulated in the adder (see the feedback loop from the output of the adder back to its input). But the results can also be passed back to the RAM directly possibly after some clipping operation.

A controller, a sample of which is shown in Figure 4, activates the whole structure in Figure 3. The controller is a pipeline in itself. The pipeline registers are the program-counter "PC" and the instruction register "IR". A stack holds the return addresses for the time loop and the various nested loops of the process specification. The instructions themselves are stored in a ROM. The state of the feedback loop is transmitted to the controller by a set of flags. In principle this structure is generic and will be part of any MISTRAL architecture ever generated. However the optimizer will choose the width of the instruction bus, the depth of the return address stack and the number of the flags such that the performance is optimized.

In addition to the optimization of the structure MISTRAL optimizes the code capturing the processes. As will be shown later optimizing the code is a prerequisite for the performance of the processor. This part of the optimization will be described in more detail in the next chapters. The techniques used are significant in a more general context and may also provide advantages for VLIW architectures.

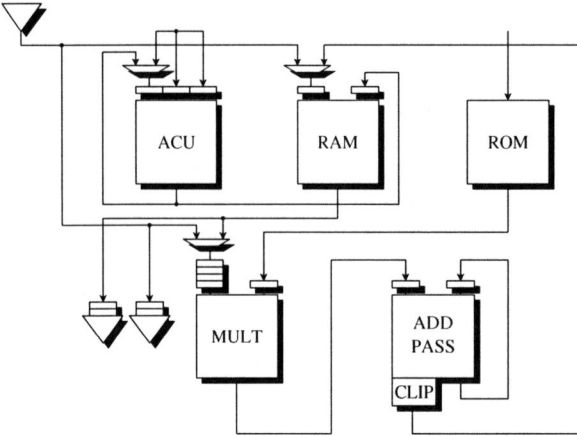

Figure 3. MISTRAL architecture adapted to application

The PHIDEO architecture is in principle built from a system of communicating parallel pipelines [MEE95]. PHIDEO processors run at moderate clock speeds. A PHIDEO processor forms the heart of the so-called I.McIC chip, which in addition also contains a MISTRAL processor ([KLE97], [WER97]). The I.McIC chip performs an MPEG2 video encoding function for applications in digital video recorders and video cameras. The encoding function implies the computation of the motion vectors and is therefore much more difficult than the decoding function. In the program code nested loops dominate the process structure. This chip contains 4.6 million transistors running at the very moderate clock frequency of 27 MHz. The PHIDEO architecture obtains its speed performance by a high degree of parallel operation combined with a highly optimized network of interconnect and (line-) memory blocks. In addition it relies on advanced techniques for the optimization

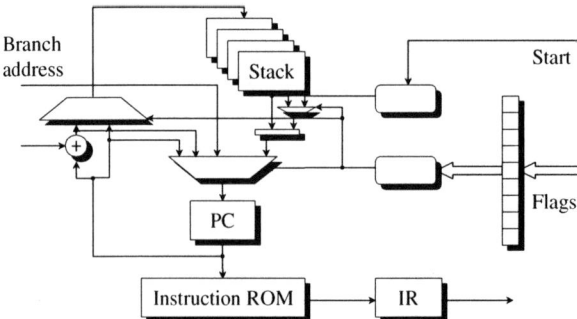

Figure 4. Controller adapted to architecture

of allocation and binding of the loop operations to various portions of the network of pipelines.

The PROPHID architecture is designed to handle a multiplicity of the media processes including those requiring massive computation like three-dimensional graphics [LEI97]. The architecture resembles the PHIDEO architecture in that it is a network of parallel operating pipelines. Both input and output of the pipelines are buffered with stacks. A high-performance switching network provides the feedback. The controller supervising the routing of data packets through the feedback loop knows about the task level dependencies between the various parallel processes.

4.   CODE GENERATION AND SCHEDULING

In this chapter we turn to a special subject namely the code generation and scheduling problems associated with media processors of the kind discussed in Chapter 3. In fact the optimization problem we address can be seen as an extension of the classical problem of code generation in a standard compiler. However the processor architectures are now different in that much more parallel operation has to be taken care of. Also as we deal with embedded systems in most instances the compiler can afford to spend more time on the code optimization. This is fairly essential for instance with the applications of the MISTRAL architecture. At times gaining one cycle in the execution sequence by better scheduling or binding makes all the difference.

4.1   The Overall Compilation Process

The overall view of the compilation process is displayed in Figure 5. The input to the compilation procedure is a document capturing the processes to be implemented by the code eventually generated. Usually this document is written in some programming language to begin with. For the purpose of this discussion, however, we assume that some preprocessing has taken place yielding a data flow graph FG_{OP} exposing the data dependencies explicitly. The details of the definition of this graph are one of the subjects of paragraph 4.2.

In Figure 5 we distinguish two iterations. The first includes the boxes "Binder" and "Constraint Analysis" and is concluded by a feasibility test. The "Binder" proposes a scheme linking the vertices of the data flow graph FG_{OP} to components of the architecture. Thus for instance a multiplication is linked to one of the available multipliers. But in addition it is decided along which ways the arguments are routed to the registers holding the inputs of the multiplication. Also it is decided which register will hold the product once the computation is completed. The "Binder" will also propose the details of load- and store-operations, as the data flow graph FG_{OP} usually doesn't contain this information. In other words: the output of the "Binder" is a set of partially ordered "Register Transfers" (RT's). The RT's can be documented in terms of another data flow graph, say, FG_{RT}, being semantically equivalent to the data flow graph at the input of the iteration. Yet this graph

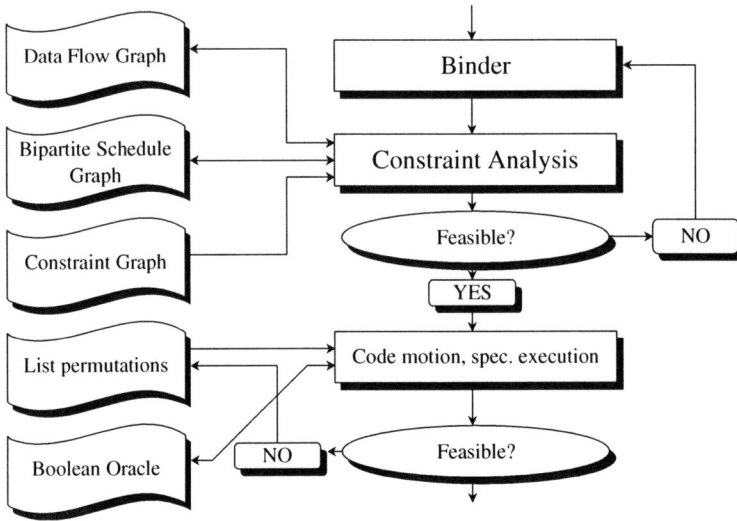

Figure 5. Overall view of the compilation process

has as vertices the RT's. The edges can be computed using the original data flow graph FG_{OP}.

The data flow graph FG_{RT} is the input to the box "Constraint Analysis". This box investigates the conflicts between the individual RT's as they compete for registers, interconnect and computational facilities. Those conflicts are captured by a "Constraint Graph" CG. The vertices of CG are taken from the set of the types of the different RT's. An edge between two vertices of a CG indicates a resource conflict between the adjacent RT-types. Such a conflict implies that instances of the respective RT's cannot share common time slots. The "Constraint Analysis" involves a scheme of forward reasoning to capture the implications of the resource conflicts. Usually this analysis leads to a narrowing of the bounds for the scheduling of the RT's. Often enough it indicates the unfeasibility of a binding scheme proposed by the "Binder" in which case backtracking is effectuated.

In case a binding scheme is found feasible the scheduling of the RT's is initiated. The scheduling procedure is in essence based on a list-scheduling algorithm. Such an algorithm requires a priority list as input. The priority list is in fact nothing but a permutation of a list of all the RT's to be scheduled. The box "Code motion, speculative execution" attempts to rearrange RT's in time using the mobility left after satisfying the data dependencies and the resource conflicts. In particular it analyses the various paths of execution as the result of conditionals. Certain RT's can be shifted forwards or backwards in time even past the conditionals they depend on to make better use the available resources by parallel processing. It may happen in such a case that a computed value is actually never used since a conditional evaluated later may lead to choosing a path other than the expected one. For this reason the execution is called "speculative". The whole procedure is known under the term "code motion" and is also applied in classical compilers ([FIS81], [NIC85]). The legality of the move of some RT depends on the logic structure imposed by the conditional predicates. A box called "Boolean Oracle" is addressed to answer the queries regarding the legality of the move under investigation [BER95].

To obtain a satisfactory schedule a variety of priority-lists may have to be tested. Those lists are generated and maintained by the box called "List permutations". The actual way to handle the lists may depend on the application. The priorities may be derived from such application. But also evolutionary schemes may be applied [HEI95].

4.2   Data Flow Graphs and Conflict Graphs

Data flow graphs come in a large variety of different syntactic formats and also with varying semantics. Most of the syntactic and semantic details are of no real significance for our purposes. The only objective we have is to capture the data dependencies between a set of operations (in the case of FG_{OP}) or between RT's (in the case of FG_{RT}).

A data flow graph is defined by two sets V and E. For simplicity we concentrate on the case of RT's. The set V is the set of vertices and their members map one to one onto the set of RT's. The set E is the set of edges and is defined over the set of pairs from V. The pairs are annotated like (R_1, R_2) for two RT's R_1, R_2. They have a first and a second element, such that the arrow of the associated edge points from R_1 to R_2. The edge means that values are produced by R_1 which are consumed by R_2. Therefore an edge constitutes a causal relationship between individual RT's. If R_1 is the origin of the some edge it must precede R_2 (pointed to by the same edge)

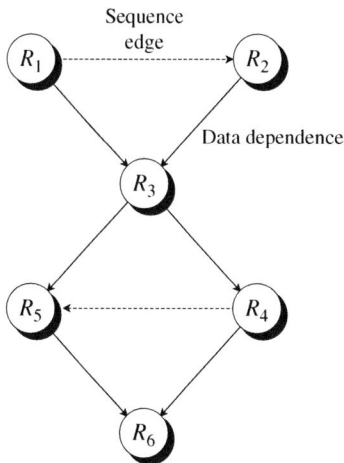

Figure 6. Example of a data flow graph

in time. In fact we assume non-preemptive scheduling. This implies that R_1 must be fully completed before the execution of R_2 can commence.

Scheduling may introduce more of these causal relations which only indirectly may be caused by data dependence. In case we want to annotate those explicitly we use "sequence edges". Figure 6 shows a sample data flow graph. Some edges are indicated as sequence edges by broken lines. Solid lines indicate the edges belonging to the original edge set E [EIJ92].

Resource conflicts preventing RT's from being executed simultaneously are captured in another graph, the so-called conflict graph CG. The vertex set of a CG is the set of "RT-types". RT's are instances of their respective types. Examples of resource conflicts are the case of two additions bound to the same adder, data transfers bound to the same bus, different RT's using the same registers or two additions with identical arguments but using the adder in a different mode. But also limitations imposed by features of the controller may introduce conflicts. For instance, considering the MISTRAL controller in Figure 4 the limitations of the instruction-width, the depth of the return address stack or the limited number of data path flags may be responsible for conflicts. In the graph conflicts are indicated by undirected edges between conflicting vertices.

For certain scheduling methods the notion of a "module" will prove to be of crucial significance (see paragraph 4.3). In the case of operations (where no "binding" has yet occurred) modules are intuitively defined. Given some operation a "module type" is simply a hardware structure able to execute it. A module is an instance of a module type. So the binding of operations to modules is a design decision only if there is more than one module of the same module type. If in some time-slot all available modules with the same module type are occupied no more of the operations of the associated kind can be scheduled in this time-slot.

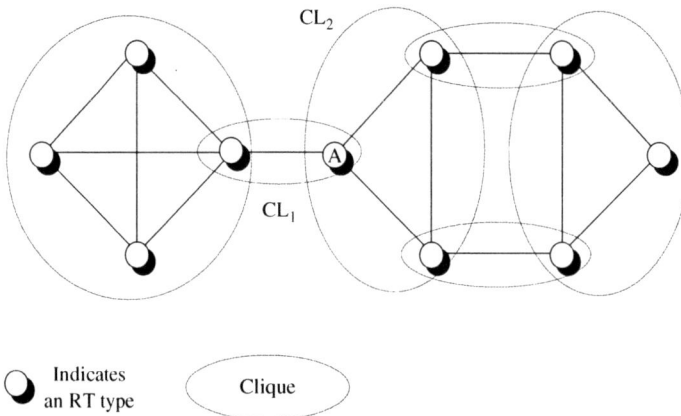

Figure 7. Example of conflict graph

In the case of RT's the binding is already done. Yet the conflict situation for scheduling when the techniques of paragraph 4.3 are applied is more difficult to oversee. There is, namely, no intuitive notion of a module. However an artificial notion of a module can be introduced as follows. Consider a complete subgraph of a CG (that is a subgraph such that any two vertices in that subgraph are mutual neighbors). Such a subgraph is also called a "clique". A clique obviously denotes a set of RT-types such that none of the RT-types in this set can be scheduled simultaneously. So such a clique may be addressed as a "module". Given a CG we can ask how many of those artificial modules are induced by the graph topology. For this purpose we consider a set of cliques covering all edges of the CG. Such a set is called an "edge covering clique". We require that all the cliques are "maximal". That is to say, no clique is a subgraph of some other clique in the cover. An example of a valid edge cover is given in Figure 7. The cliques of the cover are indicated by the contours with broken lines. We observe that for instance RT-type "A" belongs to two different cliques. This means that both involved artificial modules (indicated as CL_1 and CL_2) have to be activated simultaneously to execute an RT of type "A". Obviously the best situation is obtained if the CG decomposes into a set of mutually disconnected components. These components constitute the natural parallel operating entities of the architecture. Consequently the CG also presents a message to the designer of the architecture. If the cliques intersect a great deal the architecture gives little opportunity for parallel operation. So if even the best schedules don't yield the required performance reviewing the CG may indicate the remedies [TIM95].

The actual use of the CG in the scheduling process will be exposed more clearly in paragraph 4.3. It is, however, a complicated issue that cannot be exhaustively dealt with in this paper.

4.3 Bipartite Schedule Graphs and Interval Analysis

Given a data flow graph FG, defined by its vertex set V and its edge set E, we assign uniquely an integer from the interval $[1,|V|]$, called "name", to any vertex $v \in V$, by means of a function referred to as "name(v)". A schedule, σ, assigns a start instance $t \in [1,\omega]$ to any vertex $v \in V$ such that $\sigma(v) = t$. These definitions imply the time axis to be partitioned into a contiguous ordered set of time-slots identified by the integers of the interval $[1,\omega]$. We agree the assignment "$\sigma(v) = t$" to imply that the execution of "v" starts at the beginning of time-slot "t". The value of "ω" is addressed as the "cycle-budget" of the scheduling problem under consideration. Any vertex needs a fixed number of cycles to be executed. Note that this means a simplification of the scheduling problem in the sense that all modules eligible to perform operation "v" need the same number of cycles to execute. Our examples will show that the theory we are presenting is not restricted to this case. We accept it for this discussion for the sake of notational simplicity. The number of cycles needed to execute "v" will be denoted by $\delta(v)$. An implication of these conventions is that the execution of operation "v" ends at the end of time-slot $\sigma(v) + \delta(v)$-1. A schedule σ is legal if and only if:

(1) $\displaystyle\bigvee_{v \in V} ([\sigma(v), \sigma(v) + \delta(v) - 1] \subseteq [1, \omega])$

Otherwise the schedule is considered to be undefined. Consider a module type, say, "m". Further consider V_m to be the set of operations executable by module type "m". We define another function, "number", over the Cartesian product of the set of integers constrained to the interval $[1, |V_m|]$ and the set of schedules, by the following predicate:

(2)
$$\bigvee_{v_1, v_2 \in V_m} (\text{number}(v_1, \sigma) < \text{number}(v_2, \sigma)) \Leftrightarrow (\sigma(v_1) < \sigma(v_2))$$
$$\vee\, (\sigma(v_1) = \sigma(v_2) \wedge \text{name}(v_1) < \text{name}(v_2))$$

Any numbering satisfying this predicate assigns numbers to the operations according to their start instances determined by the schedule σ. Where σ assigns equal start instances to two or more operations "name" breaks the tie. Assume now we select an arbitrary value "i" such that $i \in \{1, \ldots, |V_m|\}$, then any schedule σ defines some operation as being the i-th operation of that schedule.

Now we define two sets of intervals, namely the "Operation Execution Intervals" (OEI's) and the "Module Execution Intervals" (MEI's). (Note that here an "interval" is a subset of contiguous time-slots from $[1, \omega]$.) Given $i \in \{1, \ldots, |V|\}$ we define OEI(i) as follows:

(3) $\text{OEI}(i) = \displaystyle\bigcup_{\sigma} \{[\sigma(v), \sigma(v) + \delta(v) - 1] | \text{name}(v) = i\}$

So the OEI is in fact an attribute of each operation. Any schedule will position the operation "v"(with name "i") into a contiguous set of intervals (with cardinality $|\delta(v)|$). The OEI associated with this operation is simply the union of those intervals taken over all legal schedules.

Let us now consider the definition of the MEI's. First we select a module type, say, "m". From the set V of operations we restrict our analysis to the set $V_m \subseteq V$ executable by module type "m". Now we select some $i \in \{1, \ldots, |V_m|\}$. Then by definition:

(4) $\text{MEI}(i, \text{m}) = \displaystyle\bigcup_{\sigma} \{[\sigma(v), \sigma(v) + \delta(v) - 1] | v \in V_m \wedge \text{number}(v, \sigma) = i\}$

To define all possible MEI's the definition has to be instantiated for each module type.

It is obvious that the notion of a MEI is in a way volatile. An MEI is not bound to a certain operation the way an OEI is. Neither is it bound to a particular module of type "m". Yet it is bound to the module type. Considering all legal schedules it defines a sort of a Gantt chart for the activity of the available modules of type "m". Given some schedule σ, as we observe the activity associated with the i-th operation defined by σ it will keep a module of type "m" busy during the respective execution interval. Uniting all these intervals bound to the i-th operation induced as such by σ establishes the fact that there must be activity of some module of type "m" in some time-slots in this MEI. Observe that the number of MEI's associated with some module type "m" is equal to $|V_m|$.

We now define a set of "m" bipartite graphs, the so-called "Bipartite Schedule Graphs" (BSG's), as follows: given "m" we obtain $BSG(m) = (O_m \cup M_m, U_m)$ where the vertex and edge sets are as defined below:

(5) $O_m = \{OEI(i) | v \in V_m \wedge name(v) = i\}$

(6) $M_m = \{MEI(i, m) | i \in \{1, .., |V_m|\}\}$

(7) $U_m = \{(OEI(i) \in O_m, \Delta \in M_m) | |OEI(i) \cap \Delta| \geq \delta(v | name(v) = i)\}$

As usual with a bipartite graph the vertex set dissolves into two disjoint sets. The vertex set O_m consists of the OEI's associated with the operations handled by module type "m". The vertex set M_m collects the MEI's associated with module type "m". Note that both vertex sets have identical cardinality. Eq. (7) defines the edges of the graph joining OEI's and MEI's. Suppose "v" is the operation associated with some OEI then there is an edge between OEI and some MEI if and only if the overlap of both in terms of time-slots is at least $\delta(v)$ [TIM93].

Now the construction of the BSG's is clear there remains the question how the OEI's and the MEI's can actually be computed. To begin with the OEI's, path analysis of the data flow graph delivers a first set of estimates. For any vertex "v" in the given data flow graph the interval needed to execute is known by the function $\delta(v)$. Also we have a given cycle budget by the value of ω. Thus if there exists at least one legal schedule then for any vertex there is a largest lower bound and a smallest upper bound delimiting the interval of actual execution. (If no legal schedule exists then the intervals are not defined.) For any vertex "v" the largest lower bound is called the "As-Soon-As-Possible"-value or the "ASAP"-value of "v". Likewise the smallest upper bound is called the "As-Late-As-Possible"-value or the "ALAP"-value of "v". These intervals are easy to compute. It can be shown that for any schedule "σ" and for any vertex "v" the ALAP – ASAP interval always covers the actual execution interval.

Good estimates for the MEI's are more difficult to obtain. Instead of going into the details of the associated theory let us consider a few simple examples. Consider Figure 8. On the left side there is a data flow graph with vertices "1" to "5". The cycle budget is defined to be $\omega = 6$. There is one single module being able to execute all vertices (operations). The execution time for each operation is $\delta(1) = \ldots = \delta(5) = 1$. Straightforward analysis shows the ALAP-ASAP intervals as given in the center Gantt chart yielding estimates for the OEI's. Consider the MEI's. We call an operation "ready" if all its predecessors in the data flow graph have been executed in previous time-slots. In time-slot "1" there is a ready operation available, namely either operation "1" or operation "2". Also a module is available. Therefore there will be a schedule among the legal schedules that will assign, say, operation "1" to the first time slot. After one time-slot the execution of operation "1" will be completed, so the module is again available at the beginning of time-slot "2". In this example for any schedule there is an alternate operation ready. So there exists a schedule assigning a free operation to the module in time-slot "2". All schedules following the same policy up to now make operation "3" ready at time-slot "3"and at the same time have the module available. Consequently there is a schedule executing operation "3" in time-slot "3". Reasoning along these lines yields the lower bounds for all the MEI's. The upper bounds of all MEI's result from a similar kind of reasoning starting from the end of the interval $[1,\omega]$. The result of all this reasoning is depicted on the right side of Figure 8. It is obvious from this example that laying the MEI's on top of the OEI's reduces the freedom for the scheduler significantly. This is exactly what is pictured in the BSG. The BSG for the example of Figure 8 is displayed in Figure 9. The reader is invited to verify the definition as given in eq. (7).

Some theoretical properties of bipartite graphs can be taken advantage of to support our scheduling efforts. First of all we can look for a so-called "matching" in the BSG ([DUL63], [HOP73]). A "matching" in a bipartite graph is a set of edges such that no two edges share any of their incident vertices. Vertices incident to edges of a matching are addressed as "covered". A "maximum matching" is a matching with maximum cardinality. A matching covering all vertices is a "complete matching". Obviously a complete matching is maximum. A complete matching can only exist if $|O| = |M|$ (see Figure 9). The BSG's satisfy this property by construction.

Every (legal) schedule corresponds with a maximum matching. Namely, every legal schedule σ defines a "number"-function (see eq. (2)) assigning a unique number to any operation. This number identifies the MEI the operation is placed into by the schedule σ (see eq. (4)). Thus a schedule σ selects a set of edges from the BSG establishing a unique pair-wise binding of OEI's to MEI's, this way defining a complete matching.

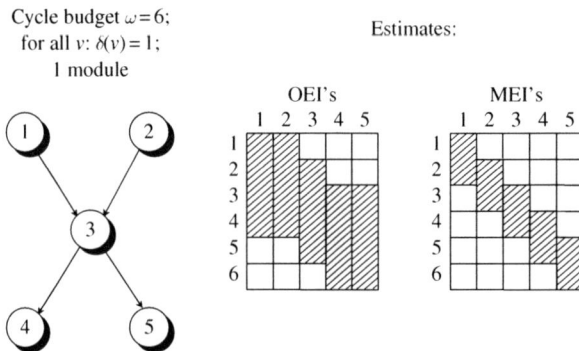

Figure 8. Data flow graph and intervals

From this statement we can conclude that the absence of a complete matching indicates the nonexistence of a legal schedule. Therefore it makes sense to initiate scheduling by searching for a complete matching in the first place. If such a matching is found a legal schedule may exist. However, this fact has to be established separately. In the example of Figure 9 a complete matching is easily identified (see for instance the emphasized edges). Moreover the indicated matching suggests two legal schedules both of length five, which are distinguished from each other by an offset of one time-slot.

Closer inspection of the edges of the BSG reveals possibilities to reduce the BSG. Consider for instance the edges (OEI(1), MEI(3)) and (OEI(3), MEI(1)). Both edges may be incorporated in an alternate complete matching. This would, however, imply operation "3" to be scheduled latest in time-slot "2". But there is only one module available implying that at least two time-slots will elapse before operation "3" is ready. Thus the alternate matching wouldn't yield a legal schedule and in fact edge (OEI(1), MEI(3)) can be deleted from the BSG. By similar reasoning many edges can be removed.

The observation that no operation can be scheduled in parallel with operation "3" binds this operation uniquely to MEI(3). Operation "4" and "5" are successors of "3" and are therefore confined to MEI(4) and MEI(5). This procedure can be generalized in the sense that, for any operation, counting the transitive set of predecessors and successors may allow to bind operations closer to MEI's thereby cutting away edges. Figure 10 shows the BSG of Figure 9 with as many edges as possible removed after transitive predecessor and successor analysis [TIM95a].

The example shows edge pruning to be the primary technique yielding the BSG a useful instrument to assist scheduling. The theory of bipartite graphs supplies further techniques. Bipartite graphs can be decomposed into their so-called "irreducible components" (which bear a resemblance to the so-called doubly connected components of undirected graphs) ([DUL63], [SAN76]). There are algorithms with a complexity linear in the number of vertices of the BSG performing the decomposition. From the theory we obtain more indications to prune the BSG. Obviously the information regarding legal schedules is one to one associated with the matchings of

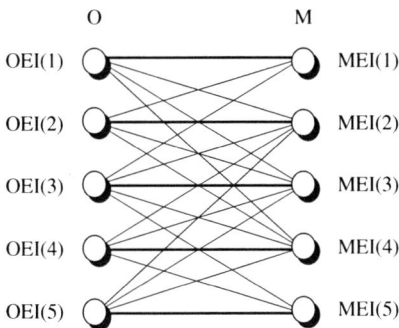

Figure 9. Bipartite Schedule Graph (BSG) for the example of Figure 8

the BSG. The theory states that edges not being element of an irreducible component can never be an element of any matching. Therefore those edges can be removed without losing any legal schedules. Without going into the details of the theory and the algorithms for the decomposition we state that the BSG in Figure 9 cannot be decomposed into any smaller irreducible components. After pruning this can very well be the case. The BSG in Figure 10 decomposes into three irreducible components, which incidentally in this case are identical with the connected components of the BSG.

Interval analysis by means of BSG's has been successfully applied in a number of industrially relevant situations in particular with the MISTRAL architecture. A small set of examples is given by way of illustration in Table 1 and Table 2.

Table 1 shows the case of a DSP application, a "Wave Digital Elliptic Filter" (WDELF) [DEW85]. The specification of the filter process is actually given in terms of a data flow graph that is not presented explicitly in this paper for brevity. The filter process involves 26 additions/subtractions of pairs of signal samples and 8 multiplications of signal samples with filter coefficients in addition to various linking operations with the outside world of the filter.

The application of interval analysis to this example implies some generalization. The function δ defining the number time-slots for some operation to execute is now defined over a domain of two sets, namely the set of operations (as before!) and the set of module types. This is to accommodate the situation that an operation can be serviced by more than one module type with possibly different execution times. In our example we allow two types of adders: a "fast adder" (performing an addition in one time-slot) and a "slow adder" which needs two time-slots to do the same. The understanding is that the slow adder is smaller and may need less power than the fast adder. The construction of the BSG's has to account for the fact that OEI's connected to additions will appear in two BSG's, each being associated with one of the two adder types. Pruning can take its normal course hoping that at one stage of pruning a decision in favor of one of the possible choices is suggested.

In Table 1 a number of different scenarios is presented. Every row starts with the specification of the cycle budget in the first column. The next three columns indicate the available modules of each type. The fifth column indicates the average "mobility" left for the operations to move over time-slots before and after the application of the interval analysis. Initially each operation may move within its ASAP-ALAP interval (given as an estimate for its OEI) less the number of time-slots for its execution. This interval denotes the "mobility". The mobility is given in counts of time-slots. Operations executable by more than one module type own a separate mobility count for each module type. Column five gives in parentheses the initial average mobility over all operations. Left next to this value the average mobility after the pruning of the BSG's is shown.

The pruning process may go as far as deciding the final positioning of some operation. The sixth column of Table 1 presents the relative count of fixed operations (given in %) with mobility "zero" after and (in parentheses) before pruning. Quite

O   M

OEI(1) ◯━━━━━━━━━━━━◯ MEI(1)

OEI(2) ◯━━━━━━━━━━━━◯ MEI(2)

OEI(3) ◯━━━━━━━━━━━━◯ MEI(3)

OEI(4) ◯━━━━━━━━━━━━◯ MEI(4)

OEI(5) ◯━━━━━━━━━━━━◯ MEI(5)

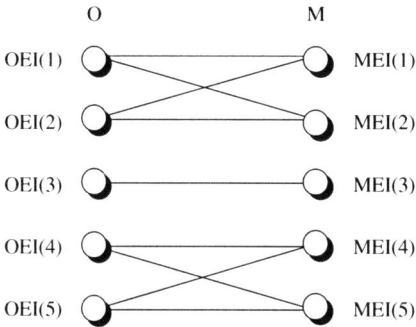

Figure 10. BSG from Figure 9 after removing redundant edges

Table 1. Results of the interval analysis for the "Wave Digital Elliptic Filter" example (WDELF)

ω	#multi $\delta = 2$	#fast-add $\delta = 1$	#slow-add $\delta = 2$	Average mobility	Fixed Operations
17	3	3	-	.38 (.82)	88.24% (70.59%)
18	2	2	-	1.38 (1.82)	32.35% (0%)
21	1	1	1	1.53 (4.82)	22.88% (0%)
28	1	1	-	9.47 (11.82)	2.94% (0%)
54	1	-	1	22.12 (37.82)	2.94% (0%)

obviously the numbers in column five and six of Table 1 give an indication of the success of the pruning technique for this example.

The results show fairly clearly that the technique helps considerably for the case of small cycle budgets and a fair amount of parallel operation. Those cases are documented in the first two rows of Table 1. The technique is less successful in the alternate cases of Table 1. But the results do show that the technique is able to perform well in severely constrained cases and those are the cases where other scheduling methods are less successful.

We continue by discussing another example, this time directly relating to MIS-TRAL. In this example we consider RT's rather than operations. Thus we consider scheduling after binding. This scheduling problem is usually much more constrained than the scheduling before binding. The pruning technique may do even better in this case. We consider two pieces of straight-line code that are derived from sig-nal processing applications in the audio domain. Artificial modules are defined by using the conflict graph model of Figure 7. Also we consider the case of "folding" the code allowing for the better exploitation of parallel processing available in the architecture. The folded code of course involves pipelined execution requiring extra registers. The analysis was slightly modified to compute the smallest cycle budget given the architecture. Again various generalizations of the BSG-concept were necessary that space doesn't permit us to describe.

Table 2. Results of interval (BSG-) analysis for two MISTRAL applications

Examples	# cliques	Stand. Sched. (time-slots)	Impr. Sched. (time-slots)
Sym. FIR & Bass B. (288 RT's unfolded)	22	43	38
Same folded	22	36	29
Port. Audio (358 RT's unfolded)	21	67	62
Same folded	21	61	58

The application is very sensitive to meeting the cycle budget constraint. Going beyond the budget is unacceptable and incurs redesign of the binding or even the architecture. In Table 2 four instances of a RT-scheduling problem are considered. The results from the BSG analysis are compared with the results of the standard MISTRAL scheduler (a heuristic list scheduler). Table 2 lists in column two the number of edge covering cliques found in the conflict graph (see Figure 7). Column three lists the optimum time-slot count for the standard scheduler while column four presents the counts obtained by interval (BSG-) analysis. The BSG analysis always yielded the smallest possible cycle budget. In the laboratory this removed doubts about the feasibility of the binding and saved a lot of redesign effort. In the "Sym. FIR & Bass Boost"-example the upper bound on the cycle budget was 32 time-slots. The analysis shows clearly that folding is compulsory unless other measures are taken. The "Portable Audio"-application has a lower bound for the cycle budget of 64 time-slots. The results indicate clearly that folding is not necessary while the heuristic scheduler suggests the contrary.

The examples illustrate the advantages of interval analysis in some cases. It should be noted that no execution times have been given. They are in the order of not more than a two minutes (measured on an HP-7000/735) for the largest examples. Although the concepts and the algorithms may take quite some reflection to be understood well the computational complexity of interval analysis is low and the methods scale reasonably for higher technological demands.

4.4   Constraint Propagation and Loop-Folding

The techniques presented below assume binding to have taken place. So generally we will talk about "RT's", and the data flow graph we consider is of the type FG_{RT}. The edges in the graph FG_{RT} not only indicate data dependencies. Rather each edge can be associated with a register holding a value. The source of the edge indicates the "producer" of the value while the "sink" indicates one of its "consumers". For a stable execution of the instructions a register is generally committed to hold its value for at least a minimal number of time slots. (If necessary an edge in the graph FG_{RT} will be annotated with that number, referred to as the "edge delay" in what follows.) Scheduling can therefore also be seen as arranging the hold-times of the registers over the range of time-slots $[1,\omega]$ (defining the cycle budget). The notion of the length of some path in an FG_{RT} is therefore given by the sum of the edge delays rather than by the sum of the execution times as provided by the δ-functions for each operation (which can be interpreted as "vertex delays"). An edge delay of "1" allows source- and sink-RT of that edge to be executed in subsequent time-slots. An edge with delay "n" requires n-1 time-slots to elapse between the end of the source-RT and the beginning of the sink-RT. For some non-adjacent pair of RT's the sum of the edge delays along the longest path between those RT's defines the minimum number of time-slots to elapse between their execution. This information can be used to improve the scheduling process.

Figure 11 displays various scenarios leading to a resolution of possible register conflicts. Consider for instance scenario (a) in Figure 11. We see altogether four RT's labeled P_1, P_2, C_1 and C_2. P_1 produces a value that is consumed by C_1. Similarly the value produced by P_2 is consumed by C_2. Both values are assumed to be different but the binder decided them to go into the same register. As the respective values are different and go into the same register they cannot be scheduled into the same time-slot. So inevitably one of the two is going to be delayed in any legal schedule such that the life times of the values do not overlap. This can be taken care of in the graph FG_{RT} by adding sequence edges. In the case of Figure 11a the sequence edge is arbitrarily inserted between C_1 and P_2. This implies that in every schedule derived from this graph priority will be given to the sequence P_1, C_1 over the sequence P_2, C_2. In this case, however, the choice is actually not motivated and the alternate choice, namely inserting the sequence edge between C_2 and P_1, could prove to be better. Suppose, however, a path of positive length to be present between C_1 and C_2. In that case P_2 has to be delayed till after the completion of C_1. So the choice is uniquely resolved by inserting a sequence edge between the latter two vertices. The situation in Figure 11c assumes a path of positive length to be present between P_1 and P_2. Again P_2 has to be delayed till after the completion of C_1 which is expressed by the sequence edge C_1, P_2.

It is not too difficult to identify more of those rules. They can be used to insert additional sequence edges. Moreover the addition of one sequence edge may induce other sequence edges by the transitive propagation of sequence constraints.

The computational procedure to effectuate the propagation requires the computation of the longest path between any pair of vertices in the graph. The result is stored in a distance matrix. The initial computation of the matrix entries can be expensive for large graphs. It needs to compute $|V|^2$ entries for a graph with vertex set V. For the updating of the matrix as more and more sequence edges are inserted many shortcuts are available. This matter needs, however, more research in the future.

Figure 12 demonstrates the effect with an example involving four different types of RT's related to four different components in a sample data path architecture. The components are addressed as "ROMCTRL", "ACU", "RAM" and "ALU". After applying propagation many sequence edges are inserted (they are indicated by broken lines!).

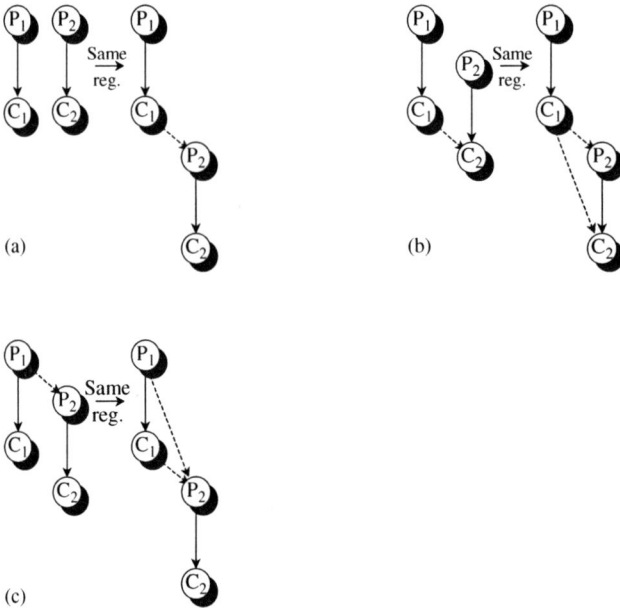

Figure 11. Resolution of register conflicts

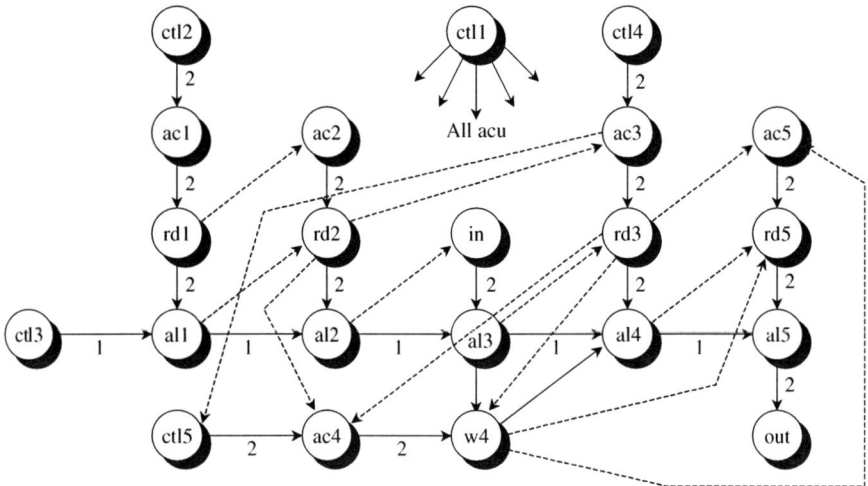

Figure 12. Illustrating constraint propagation

The procedure implies a reduction of the mobility of the RT's in the example very much as in the case of interval analysis. Some quantitative impression is given in Figure 13. Here the Gantt charts for the four different RT-types are drawn. The cycle budget is assumed to be ten time-slots.

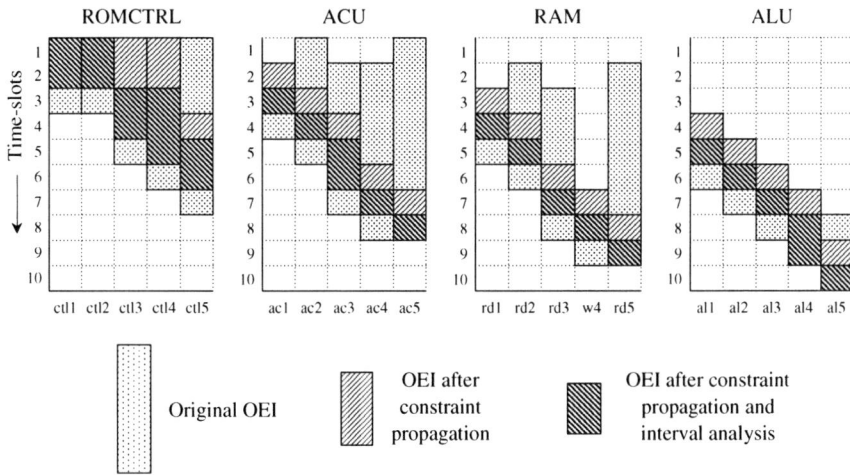

Figure 13. Narrowing of the OEI's by constraint propagation and interval analysis

As can be observed from Figure 13 the reduction of the OEI's are in certain cases considerable. Also it is evident that cascading various forms of analysis is beneficial. Combining interval analysis with constraint propagation yields still more reductions of mobility.

Constraint propagation can be modified to accommodate a technique generally known as "Loop-Folding". This notion is illustrated in Figure 14. A loop in a program consists of a loop-body (a piece of straight-line code) and a conditional telling how often the loop-body has to be executed. In a more general setting the loop-body may contain more ("nested") loops. For ease we do not consider this generalization.

On execution the loop-body is instantiated a number of times in compliance with the conditional. Any instance of the body creates in principle its own local values. The idea of Loop-Folding is to establish an overlay of different instances of the loop-body. The execution of the various loop-bodies in parallel allows for better exploitation of the available hardware.

Consider Figure 14. In the first place we see three instances of the loop-body which are numbered appropriately to distinguish between them. The schedule is supposed to overlay complementary portions of a number of loop-bodies such that all these portions are simultaneously executed. In general before scheduling a number of parameters have to fixed. In the example of Figure 14 for instance we start by fixing the data introduction interval "dii" to be three time-slots. Indeed the execution of the loop is pipelined, explaining why this technique is also called "Loop-Pipelining". Furthermore we decide to admit a latency "l" of six time-slots. In other words one instance of the loop-body is completed in six time-slots. As a consequence the folding degree "f" assumes the value "2" meaning that two instances of the loop-body are executed in parallel.

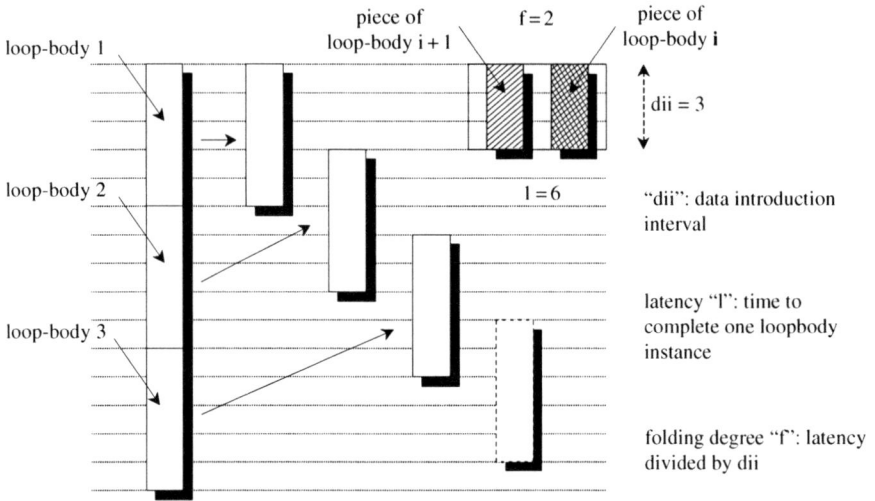

Figure 14. Introducing "Loop-Folding"

We are now ready to consider scheduling the whole. The RT's in the loop-body must be assigned to the entries of a matrix comprising "f" columns and "dii" rows. Each column will hold RT's from different instances of the loop-body. The rows are one to one correlated with "dii" different so-called "time-potentials". The following rules have to be obeyed:

– The sequence of RT's must respect the data dependencies both within one instance of the loop-body and between different instances of the loop-body;
– The union of RT's over all columns must equal the whole loop-body;
– There must be no resource conflicts between RT's in one row (the same time-potential);
– The life-span of values assigned to one register must not overlap in terms of time-potentials.

The last rule implies that the sum of the life-spans of all values assigned to one register must never exceed the value of dii. This way the pipelining may cost extra registers compared to the non-pipelined case.

The next example will show that constraint propagation can be effectively applied to narrow the search space for the scheduler in the case of Loop-Folding. Consider Figure 15a. We assume a loop-body with five RT's "A", "B", "C", "D" and "E". Resource conflicts are assumed to exist pair-wise between "A" and "D" and "B" and "D". Furthermore "dii" is chosen to be "3" and the latency "l" is chosen to be "6". A heuristic "greedy" scheduler might arrange "A", "B" and "C" to the first three time potentials in the first column. That leaves "D" to be scheduled in the first time-potential in the second column. This is not permitted because of the conflict "A" – "D". In moving down "D" we encounter the conflict "B" – "D" and therefore "D" moves down to the third time-potential. As a consequence "E" moves

out of the matrix yielding this schedule illegal. Is there a legal schedule? Yes, but
then the execution of "B" must be delayed for one time-potential down to the third
row leaving an empty spot in the matrix. This leaves room for scheduling "C",
"D" and "E" in this order in the second column avoiding all the conflicts. Below
we will demonstrate that by constraint propagation a "greedy" scheduler may be
conditioned to delay "B" for one time-potential, which is necessary to obtain the
optimal schedule.

Note first the way the latency constraint is incorporated as a sequence edge.
This is the edge from sink to source with delay "-6". Remember that the delay
is the minimum distance in time slots between two RT's. A minimum distance
from sink to source of "-6" time-slots translates into a maximum distance of "6"
time-slots between source and sink which is what we want. To trace conflicts we
have to compute minimum distances modulo "dii". This way we find out about
RT's ending up in the same time-potential if scheduled at minimum distance. For
instance the minimum distance module "dii" of the pair "A" – "D" is "0". Therefore
the minimum distance "A" – "D" must be increased by at least "1". This can be
taken care of by a sequence edge with delay "4" originating from "A" and ending
in "D" (see Figure 15b). Consider now the path "D"-"E"-"sink"-"source"-"A"-"B".
This is the longest path between "D" and "B". The minimum distance is the delay
along this path which is "-3" which modulo "dii" evaluates to "0". Because of the
conflict "D" – "B" the latter has to be shifted one row down. This is documented
in Figure 15c by the insertion of a sequence edge from "D" to "B" with delay
"-2". By "forward reasoning" we can now infer that the distance between "A"
and "B" must be "2" or greater, as indicated by an additional sequence edge from

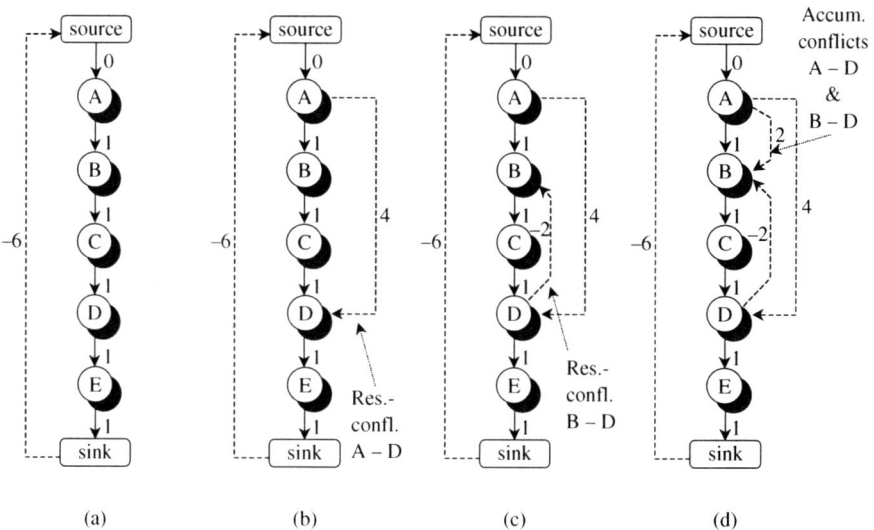

Figure 15. Constraint propagation with Loop-Folding

Table 3. Iterating register bindings: some results

Trial	#op's	dii	Lat: "I"	#iter.	Run-time	Mob. before	Mob. After
IIR	23	6	10	3	.2s	2.70	.13
FFTa	44	4	13	11	17s	4.46	.46
FFTb	60	8	18	20	25s	6.85	.52
Rad4	81	4	11	1	.8s	4.93	1.38

"A" to "B". This forces every subsequent scheduling action to postpone "B" by one time-potential, which is the key to finding the legal schedule for this case [MES97].

We conclude this discussion by presenting scheduling results concerning some larger examples illustrating the potential of constraint propagation in the case of Loop-Folding. The examples refer to instruction sequences typical for signal processing like an "Infinite Impulse Response Filter" (IIR-Filter), various forms of discrete Fourier Transforms (FFT's) and a "Radix-4 Butterfly". The experiment involves iterating a series of register binding proposals in order to find a schedule satisfying a given latency constraint. Analyzing the loops in the data flow graph controls the iteration. Positive weight loops indicate illegal schedules. A path between two instructions involving a positive loop makes the longest path between these instructions exhibit infinite delay. In choosing other bindings using the loop information the designer is often able to resolve the scheduling problem within a few iterations.

The results of the experiments can be studied in Table 3. Column one lists the names of the code segments tried. Column two gives the number of instructions in the loop-bodies considered. Column three and four give the data introduction interval (dii) and the latency constraint (l) respectively. Column five indicates the number of iterations for a trained designer to obtain legal schedules. The run-times given in column six include the CPU-times for all iterations added up. Column seven and eight display the average mobility over all operations before and after the iteration. It can be observed that the reduction of the mobility obtained by constraint propagation is very significant, certainly if measured against the CPU-times consumed [MES98].

4.5   Code Motion and Speculative Execution

"Code Motion" (in the sequel referred to as "CM") and "Speculative Execution" (further referred to as "SE") are techniques applied in connection with conditionals. CM refers to the fact that operations, instructions or RT's can be moved within their respective OEI's, for as much as data dependencies and resource conflicts permit.

In the presence of conditionals code can be moved through the conditionals. In such a case operations may be scheduled for execution at time-slots where the value of the conditional is yet unknown and therefore the result may later have to be discarded. For this reason the term "Speculative Execution" has been coined.

The potential benefits of CM and SE are in the possibly improved exploitation of the parallel operation within the processor. There are, however, also limitations. In many processors the controller limits the way and the speed of the evaluation of the conditionals. Controllers may have to reorder or unravel the conditional predicates as their potential to deal with Boolean clauses is limited. An example is a "case-statement" with many outcomes, which may have to be unraveled into a tree of "if-then-else-statements". For CM to be effective it seems to be important that the scheduling method has control over the unraveling process. Some controllers also evaluate conditional predicates in a multi-cyclic pipelined manner. Efficient methods to schedule CM and SE will have to account for this circumstance and eventually should be able to take advantage of such architectural peculiarity. Also in cases of excessive application of SE too many operations might be executed where the results are never used with negative implications for the power consumption of the processor.

In Figure 16 the principle of SE is illustrated. The specification of the program (or process) is in terms of a data flow graph. The execution follows the usual token flow semantics [EIJ92]. As compared to the previous examples there is an additional type of vertex representing the conditionals. In fact there are "split-" and "merge-" vertices (see Figure 16). The "split-" vertex looks at the input at the side of the vertex connected to the broken line - the control input -, where it expects a token carrying a Boolean value. In consuming this token it sets the switch to the left or the right output according to some fixed convention regarding the Boolean value of the token. Subsequently the vertex is ready to rout one data token from

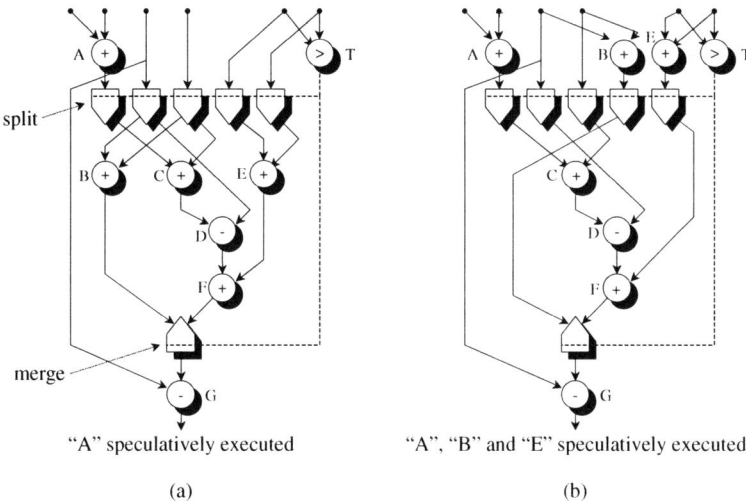

"A" speculatively executed   "A", "B" and "E" speculatively executed

(a)   (b)

Figure 16. Illustration of Speculative Execution

its input to the selected output. To pass the next data token another Boolean token is required. Note that it is possible for the "split-" vertex to rout tokens into the "empty space". This accounts for the phenomenon of removing unused data from registers.

A "merge-" vertex implements the inverse function. Depending on the Boolean value of the token at the control input it either selects the left or the right input to transfer a token residing there. The "merge-" vertices arrange for the reconvergence of different paths created by the "split-" vertices.

In Figure 16a the operation "A" can be considered as speculatively executed. In Figure 16b the operations "A", "B" and "E" are all speculatively executed. The examples suggest many choices for code to be moved around. In certain cases code may have to be copied into various paths, namely if code is moved upwards through a "merge-" vertex. "Flushing" unused values may require additional extra code. All these aspects have to be accounted for. Any optimization may consequently be quite difficult. Therefore effective pruning is an important item in the application of CM and SE.

A specific problem is the control over the execution conditions of the various paths. Some terminology will be introduced to come to grips with this matter. A "Basic Block" (BB) is a piece of code delimited by a "split-" and a "merge-" vertex. The BB's in Figure 16a are $\{A, T\}$, $\{B\}$, $\{C, E, D, F\}$ and $\{G\}$. Every BB owns an "Execution Predicate" $\mathbf{G}(BB)$, which is a Boolean predicate defined over the domain of the conditionals. For example we can derive from Figure 16a that $\mathbf{G}(\{A, T\}) = 1$, $\mathbf{G}(\{B\}) = \neg T$, $\mathbf{G}(\{C, E, D, F\}) = T$ and $\mathbf{G}(\{G\}) = 1$. If an operation moves up into another BB it inherits this BB's execution predicate. For example consider Figure 16b. After moving "B" and "E" we obtain $\mathbf{G}(\{A, B, E, T\}) = 1$, $\mathbf{G}(\{C, D, F\}) = T$ and $\mathbf{G}(\{G\}) = 1$. For any operation, say, "A", there is a "latest" BB such that "A" cannot be moved to any other BB downstream (except by moving a "merge-" vertex further downstream). BB is then called "compulsory" relative to "A". Let some operation move from BB_1 to BB_2. Then for the move to be correct $(\mathbf{G}(BB_1) \to \mathbf{G}(BB_2)) = 1$ must be valid. Let "A" move from BB_1 to BB_2 and BB_3. Then for correctness we require $(\mathbf{G}(BB_1) \to \mathbf{G}(BB_2) \vee \mathbf{G}(BB_3)) = 1$. (The symbol "$\to$" indicates the logic implication.)

The scheduling starts with a permutation Π of the operations obtained from the "List Permutation" generator (see Figure 5). Starting with time-slot "1" it steps consecutively through the time-slots. In any time-slot it considers the ready operations in the order suggested by Π and tries to assign as many of them as possible to free modules. Also at any time-slot for any BB the status of the conditionals is checked. A conditional is assigned a "don't care" if its evaluation is not completed. Otherwise it is assigned the appropriate truth-value. Thus at any time-slot, say, "c" any operation, say, "A" originally in BB owns a "Dynamic Execution Predicate", say, $\Gamma(c, A)$, which is equal to $\mathbf{G}(BB)$ with all "don't care" valued conditionals removed by existential quantification (so-called "smoothing"). Pairs of mutually exclusive operations can share a resource. Let in some time-slot "c" $A \in BB_1$ and

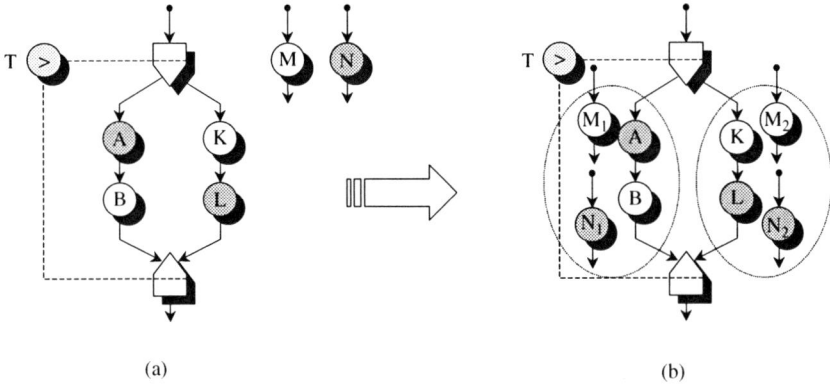

$$G(\{A,B\}) = T; \qquad G(\{M,N\}) \rightarrow \qquad G(\{A,B,M_1,N_1\}) = T;$$
$$G(\{K,L\}) = \neg T; \qquad G(\{A,B,M_1,N_1\}) \vee G(\{K,L,M_2,N_2\}) = 1 \qquad G(\{K,L,M_2,N_2\}) = \neg T;$$
$$G(\{M,N\}) = 1$$

Figure 17. Example for scheduling with CM and SE

$B \in BB_2$ then the condition $\Gamma(c, BB_1) \wedge \Gamma(c, BB_2) = 0$ ensures "A" and "B" to be mutually exclusive.

To illustrate the above consider the example in Figure 17. On the left side we see a data flow graph with a conditional "T" and two separate operations "M" and "N". There are altogether three modules, one of them being the comparator establishing the conditional. The binding of operations to modules is indicated by the patterns of the vertices. The BB's and the associated execution predicates are given in the lower left corner of Figure 17.

We suggest moving the operations "M" and "N" into the BB's of the conditional. This yields the situation of Figure 17b. Note that the operations "M" and "N" have to be replicated twice, once in each path of the body of the conditional. From the expressions below in Figure 17 (but in this simple example also by inspection!) we see that this replication is necessary because otherwise the conditions for correctness are not satisfied.

After this preparation the actual scheduling can now take place. The "List Permutations" generator is invoked. Say it supplies the permutation Π_1 (see Figure 18a). As a result "A" and "M" are moved upwards into time-slot "c1", which results in SE for "A". In principle "L" and "A" are mutually exclusive, so one could consider them to share their resource in time-slot "c1". This would shorten the total execution time to two time-slots. However, in time-slot "c1" the value of the conditional "T" is not known yet, so the sharing would yield incorrect results. An alternate permutation Π_2 delivers the schedule depicted in Figure 18b. More permutations can be explored if the designer so desires. Critical path analysis reveals that, given the resources no schedule shorter than three time-slots exists. For power consumption the best schedule would probably be to have "N" and "M" executed in time-slot

$$\Pi_1 = [A,M,B,K,L,N,T]$$ $$\Pi_2 = [K,N,A,M,B,L,T]$$

c1

c2

c3

(a)   (b)

$$\Gamma(c1,\{A,M,T\}) = \Gamma(c1,\{N,B\}) = \Gamma(c1,\{N,K,L\}) = 1;$$
$$\Gamma(c1,\{N,K,T\}) = \Gamma(c1,\{A,M,B\}) = \Gamma(c1,\{L,M\}) = 1;$$

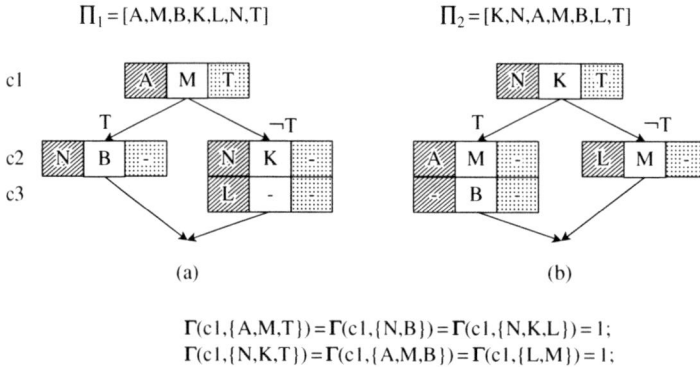

Figure 18. Two schedules for the graph in Figure 17b

"c1" because then all computed values are actually used and the total activity of the processor is minimized [SANTOS96].

The implementation of these techniques relies on efficient graph algorithms to handle the transformations of the data flow graph accommodating the moves of operations or even whole groups of code. Furthermore a very efficient "Boolean Oracle" is required able to answer the queries regarding the execution predicates quickly [RAD96]. Modern program packages handling Boolean predicates in the form of so-called "Binary Decision Diagrams" (BDD's) provide this facility [BER95].

5.   CONCLUSIONS

In summarizing we conclude the following:
- New multimedia platforms exhibit more parallel operation, lower clock-speed and more memory access features as compared to standard micro-processors. In the architectures of those platforms many special memory features (line-memories, FIFO stacks) and interconnect features (crossbar-switch) exist requiring special compilation techniques. The intention is to obtain higher performance not by up-scaling the clock-speed, but alternatively to avoid cache misses and page faults by taking advantage of the knowledge of the very special data formats present in video- and audio-processes.
- Compilers for those platforms have more time to perform optimizations. A compiler for an embedded process may still be acceptable if it runs for some hours once it succeeds to rigidly optimize the schedule of some embedded code. This is in accordance with the very strict performance control typical for embedded processes. Cycle budgets and power limits are extremely critical.
- The compilation techniques offered here use the heavily constrained character of the code scheduling problem to their advantage. "Interval Analysis", "Constraint Propagation", "Loop-Folding" and "Code Motion" all work the better the tighter the constraints are.
- Merging "Loop-Folding" with "Code Motion" still needs some new concepts.

This paper indicates a direction and tries to establish a map of the problems we are facing when designing the next generations of compilers for "systems on a chip" for media applications.

ACKNOWLEDGEMENT

A great number of people have directly or indirectly contributed to this paper. I would like to acknowledge the help of Jos van Eijndhoven, Marc Heijligers, Rick Hilderink, Wim Philipsen, Marino Strik, Leon Stok and Adwin Timmer. Leon Stok was the first to graduate for his Ph.D. in the "High Level Synthesis" area in my group. Jos, Marc, Rick, Wim and Adwin are the architects of our design environment called "New Eindhoven Architectural Toolbox" (NEAT), which has been the framework for most of our research [EIJ92]. Jeroen Leijten is the architect of the PROPHID architecture. Adwin Timmer is the originator of the "Interval Analysis" method. Marino Strik designed and implemented the MISTRAL version of the interval analysis. Bart Mesman developed the "Constraint Propagation" approach. Luiz Villar dos Santos worked on the "Code Motion" issue. Jos van Eijndhoven was also involved in developing the compilers for the TRIMEDIA processor and contributed to the evolution of its architecture. For as much as I know M. Berkelaar and L. van Ginneken coined the term "Boolean Oracle" in [BER95]. G. Janssen provided the implementation of the "Boolean Oracle" which is based on his "BDD"-package [JAN95]. Finally Koen van Eijk implemented most of the scheduling techniques in the scheduling framework FACTS [Eijk00].

Finally I am greatly indebted to Jef van Meerbergen who has supported our work in all kinds of ways in the recent years.

REFERENCES

[BER95] Berkelaar, M., van Ginneken, L., (1995) Efficient Orthonormality Testing for Synthesis with Pass-Transistor Selectors, Proceedings of the IEEE International Conference on CAD for Circuits and Systems (ICCAD), San Jose, pp. 256-263.

[DEW85] Dewilde, P., Deprettere. E., Nouta, R., (1985) Parallel and Pipelined VLSI Implementations of Signal Processing Algorithms, in S.Y. Kung, H.J. Whitehouse and T. Kailath, "VLSI and Modern Signal Processing", Prentice Hall, pp. 258-264.

[DUL63] Dulmage, A.L., Mendelsohn, N.S., (1963) Two Algorithms for Bipartite Graphs, Jour. Soc. Indust. Appl. Math., Vol. 11, No. 1. pp. 183-194.

[EIJ92] Van Eijndhoven, J.T.J., Stok, L., (1992) A Data Flow Graph Exchange Standard, Proceedings of the European Design Automation Conference (EDAC), Brussels, pp. 193-199.

[Eijk00] Eijk, van C.A.J., Mesman, B., Alba Pinto, C.A., Zhao, Q., Bekooij, M., van Meerbergen, J.L., Jess, J.A.G., (2000) Constraint Analysis for Code Generation: Basic Techniques and Applications in FACTS, ACM Transactions on Design Automation of Electronic Systems, Vol. 5, Nr. 4, pp. 774-793.

[FIS81] Fisher, J.A., (1981) Trace Scheduling: A Technique for Global Microcode Compaction, IEEE Transactions on Computers, Vol. C-30, No. 7, pp. 478-490.

[HEI94] Heijligers, M.J.M., Hilderink, H.A., Timmer, A.H., Jess, J.A.G., (1994) NEAT, an Object Oriented High-Level Synthesis Interface, Proceedings of the IEEE International Symposium on Circuits and Systems (ISCAS), London, pp. 1233-1236.

[HEI95] Heijligers, M.J.M., Cluitmans, L.J.M. Jess, J.A.G., (1995) High Level Synthesis Scheduling and Allocation Using Genetic Algorithms, Proceedings of the Asia and South Pacific Design Automation Conference, pp. 61-66.

[HOP73] Hopcroft, J.E., Karp, R.M. (1973) An $n^{5/2}$ Algorithm for Maximum Matching in Bipartite Graphs, SIAM Journal of Comp., Vol. 2, No. 4, pp. 225-231.

[HUR96] Van den Hurk, J. (1996) Hardware/Software Codesign: an Industrial Approach, Ph.D. Thesis, Eindhoven University of Technology, Eindhoven, The Netherlands.

[ISO93] International Organization for Standardization, (1993) Generic Coding of Moving Pictures and Associated Audio Information – Part 2, ISO/IECJTC1/SC29 N 659.

[ISO94] International Organization for Standardization, (1994) Generic Coding of Moving Pictures and Associated Audio: AUDIO, ISO/IEC 13818-3. ISO/IECJTC1/SC29/WG11 N0803, 11 November 1994.

[ISO94a] International Organization for Standardization, (1994) Generic Coding of Moving Pictures and Associated Audio: SYSTEMS, Recommendation ITU-T H.222.0 ISO/IEC 13818-1. ISO/IECJTC1/SC29/WG11 N0801, 13 November 1994.

[ISO95] International Organization for Standardization, (1994) Generic Coding of Moving Pictures and Associated Audio: VIDEO, Recommendation ITU-T H.262 ISO/IEC 13818-2. ISO/IECJTC1/SC29/WG11 N0982, 20 January 1995.

[ITRS] The International Technology Roadmap for Semiconductors, http://public.itrs.net

[JAN95] Janssen, G.L.J.M., (1995) Application of BDD's in Formal Verification, Proceedings of the 22nd International School and Conference, Yalta (Gurzuf), Ucraine, pp. 49-53.

[KLE97] Kleihorst, R.P., van der Werf, A., Brulls, W.H.A., Verhaegh, W.F.J., Waterlander, E., (1997) MPEG2 Video Encoding in Consumer Electronics, Journal of VLSI Signal Processing, Vol. 17, pp. 241-253.

[LEG91] LeGall, D., (1991) MPEG: A Video Compression Standard for Multimedia Applications, Communications of the ACM, Vol. 34, No. 4, pp. 46-58.

[LIP96] Lippens, P.E.R., De Loore, B.J.S., de Haan, G., Eeckhout, P., Huijgen, H., Lovning, A., McSweeney, B.T., Verstraelen, M.J.W., Pham, B., Kettenis, J., (1996) A Video Signal Processor for Motion-Compensated Field-Rate Upconversion in Consumer Television, IEEE Journal of Solid State Circuits, Vol. 31, pp. 1762-1769.

[LEI97] Leijten, J.A.J., van Meerbergen, J.L., Timmer, A.H., Jess, J.A.G., (1997) PROPHID: a Data Driven Multi-Processor Architecture for High-Performance DSP, Proceedings of the European Design and Test Conference (ED&TC), Paris, p. 611.

[MAN90] De Man, H., Catthoor, F., Goossens, G., Vanhoof, J., van Meerbergen, J.L., Huisken, J, (1990) Architecture Driven Synthesis Techniques for VLSI Implementation of DSP Algorithms, Proceedings of the IEEE, pp. 319-335.

[MEE95] Van Meerbergen, J.L., Lippens, P.E.R., Verhaegh, W.F.J., van der Werf, A., (1995) PHIDEO: High Level Synthesis for High-Throughput Applications, Journal of VLSI Signal Processing, Vol. 9, pp. 89-104.

[MES97] Mesman, B., Strik, M.T.J., Timmer, A.H., van Meerbergen, J.L., Jess, J.A.G. (1997) Constraint Analysis for DSP Code Generation, Proceedings of the ACM/IEEE International Symposium on Systems Synthesis (ISSS), Antwerp, pp. 33-40.

[MES98] Mesman, B., Strik, M.T.J., Timmer, A.H., van Meerbergen, J.L., Jess, J.A.G. (1998) A Constraint Driven Approach to Loop Pipelining and Register Binding, Proc. Conference on Design, Automation and Test in Europe, DATE, Paris, France, 23-26 February 1998, ISBN 0-8186-8361-7, ed. A. Kunzmann; IEEE Computer Society, Los Alamitos, CA, 1998, pp. 377-383.

[MES01] Mesman, B., Constraint Analysis for DSP Code Generation, Ph.D. Dissertation, Eindhoven University of Technology, May 2001.

[NIC85] Nicolau, A., (1985), Uniform Parallelism Exploitation in Ordinary Programs, Proceedings of the International Conference on Parallel Processing, pp. 614-618.

[RAD96] Radivojevic, I., Brewer, F., (1996) A New Symbolic Technique for Control-Dependent Scheduling, IEEE Transactions on CAD for Circuits and Systems, Vol. 15, No. 1, pp. 45-57.

[SAN76] Sangiovanni-Vincentelli, A., (1976) A Note on Bipartite Graphs and Pivot Selection in Sparse Matrices, IEEE Transactions on Circuits and Systems, Vol. CAS 23, No.12, pp. 817-821.

[SANTOS96] Villar dos Santos, L.C., Heijligers, M.J.M., van Eijk, C.A.J., van Eijndhoven, J.T.J., Jess, J.A.G., (1996) A Constructive Method for Exploiting Code Motion, Proceedings of the ACM/IEEE International Symposium on Systems Synthesis (ISSS), San Diego, pp. 51-56.

[SLA96] Slavenburg, G.A., Rathnam, S., Dijkstra, H., (1996) The Trimedia TM-1 PCI VLIW Media Processor, Hot Chips 8 Symposium, Stanford University, August 18-20, 1996, Stanford, California, http://infopad.EECS.Berkeley.EDU/HotChips8/6.1/

[STR95] Strik, M.T.J., van Meerbergen, J.L., Timmer, A.H., Jess, J.A.G., Note, S., (1995) Efficient Code Generation for In-House DSP-Cores, Proceedings of the European Design and Test Conference (ED&TC), Paris, pp. 244-249.

[TIM93] Timmer, A.H., Jess, J.A.G., (1993) Execution Interval Analysis under Resource Constraints, Proceedings of the IEEE International Conference on CAD for Circuits and Systems (ICCAD), Santa Clara, pp. 454-459.

[TIM95] Timmer, A.H., Strik, M.T.J., van Meerbergen, J.L., Jess, J.A.G. (1995) Conflict Modeling and Instruction Scheduling in Code Generation for In-House DSP Cores, Proceedings of the 32nd Design Automation Conference (DAC), pp. 593-598.

[TIM95a] Timmer, A.H., Jess, J.A.G., (1995) Exact Scheduling Strategies Based on Bipartite Graph Matching, Proceedings of the European Design and Test Conference (ED&TC), Paris, pp. 42-47.

[WER97] Van der Werf, A., Brulls, W.H.A., Kleihorst, R.P., Waterlander, E., Verstraelen, M.J.W., Friedrich, T., (1997) I.McIC: A single Chip MPEG2 Video Encoder for Storage, Proceedings of the International Solid State Circuits Conference (ISSCC), pp. 254-255.

[WEI96] Weiser, U., Trade-off Considerations and Performance of Intel's MMX™ Technology, (1996) Hot Chips 8 Symposium, Stanford University, August 18-20, 1996, Stanford, California, http://infopad.EECS.Berkeley.EDU/HotChips8/5.1/

CHAPTER 4

PAST, PRESENT AND FUTURE OF MICROPROCESSORS

FRANÇOIS ANCEAU

Conservatoire National des Arts et Métiers (CNAM), Paris, France
ASIM/Lip6 lab., U. Pierre & Marie Curie, Paris, France

Abstract: Microprocessors are one of the most important technical phenomena of the end of the 20th century and the beginning of the 21st century. For thirty-five years their computing power and their complexity have increased at sustained rates. Microprocessors are increasingly playing a major role in the modern society. The embedded ones are the most numerous, used for controlling and monitoring machine tools, cars, aircraft, consumer electronics and other equipments. There is a gradually changing on the relationship we have with these devices. It is interesting to show that microprocessor phenomenon is "market-pull" rather than "technology-push". The design of new chips represents a continuous challenge for engineers and technologists. They are striving to give to the market the products it requires, and which are generally planned long time before they actually appear. Monolithic microprocessors are overtaking all kinds of computers. Minicomputer lines started using microprocessors during 1980s; mainframe and Unix lines during the 1990s and super-computers during this decade. In this extraordinary evolution, these devices have used all technical innovations that had been conceived for previous generations of computers. To keep an evolutionary rate of computing power and compatibility at binary code level, completely new execution techniques has been invented as data-flow processors, register renaming, SIMD instructions and VLIW execution technique. Future of these devices is very challenging. The need for more and more computing power is still present and technology will finish by reaching physical limits

Keywords: Microprocessor architectures, microprocessor design, VLSI, microprocessors evolution

1. INTRODUCTION

The story of commercial microprocessors (that we could call VLSI processors) began in 1972 with Intel 4004 that was used in calculator and in many others applications. Since this date, microprocessors have become powerful computers

65

R. Reis et al. (eds.), Design of Systems on a Chip, 65–82.
© 2006 *Springer.*

that are taking over all other ranges of computers. Microprocessors are one of most important technical phenomena of end of 20th century and of beginning of 21st century.

2.   A SUSTAINED EVOLUTION RATE

For more than 35 years, the evolution of microprocessors is kept at a sustained rate. Complexity of these devices increases with a rate close to 37% per year and their performances (measured with a common evaluation tool) increase with a same rate.

The evolution of complexity (Figure 1) of these devices, in number of transistors, is exponential showing a constant technological evolution. The minimum feature size of technology decreases regularly (Figure 3) while size of chips increases slowly.

Everybody (including experts) is continuously predicting a decreasing of this evolution rate because it is increasing the difficulty to manufacture such devices. But real evolution has not taken into account these warnings and continues its same fast evolution rate. Currently the minimum feature size of best commercial technology is now around 65nm.

However, the evolution rate of the computing power (Figure 2) of microprocessors has decreased since 2002. The reason presented by main manufacturers is the dramatic evolution of power consumption of theses chips. Difficulties on cooling them and the move of the market toward portable computers drives the shift from high-performance processors to low power ones.

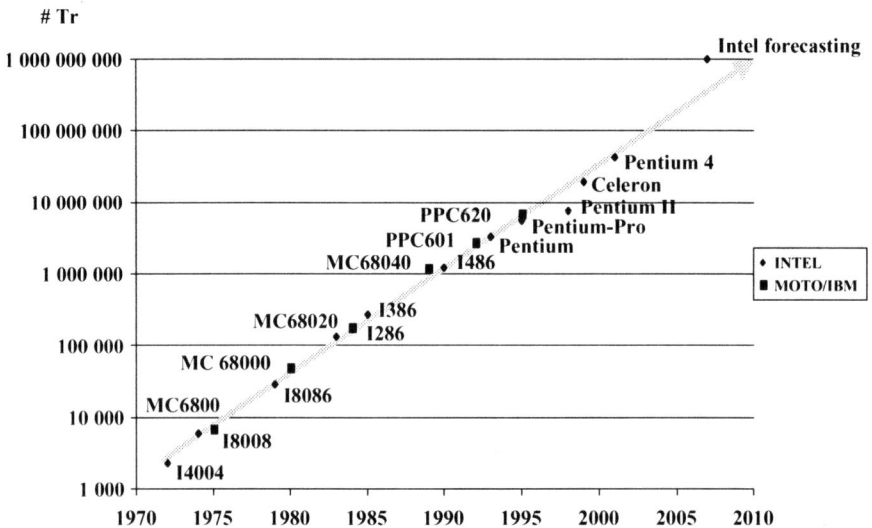

Figure 1. Evolution of microprocessors complexity

Performance (specint 2000)

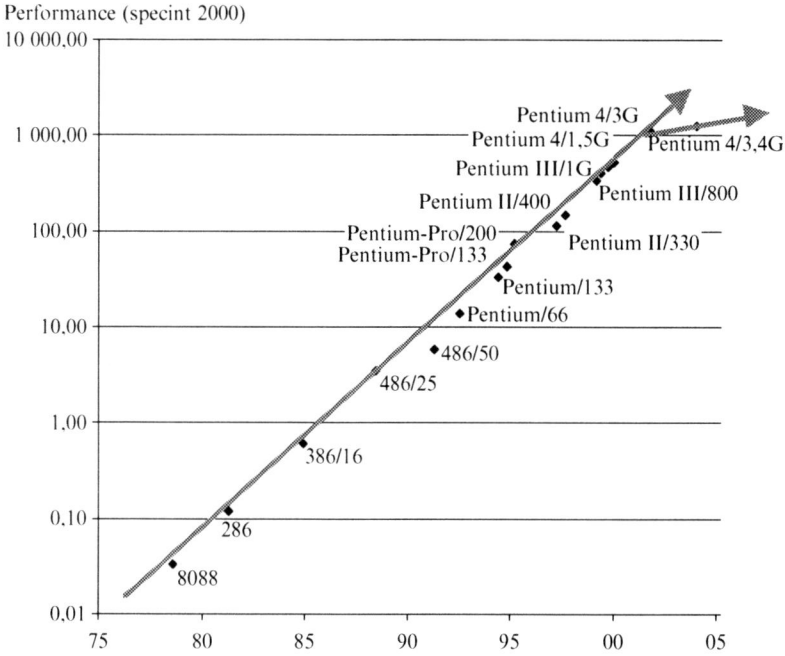

Figure 2. Evolution of microprocessors performance

Minimum feature size (microns)

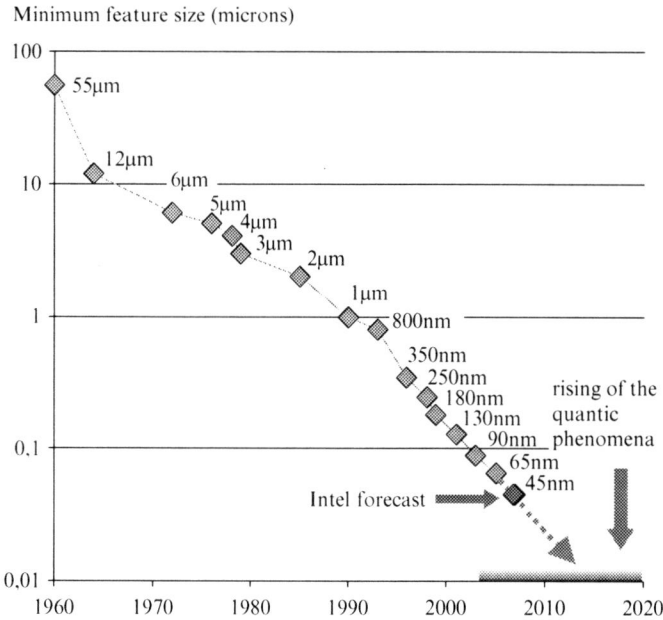

Figure 3. Technological evolution

3. VLSI PROCESSORS AS MASS-PRODUCTION COMMONALITIES

From the economical point of view, microprocessors have contributed to the democ-ratization on the access to computers. They have contributed to the decreasing price of computers by dropping the cost of their electronic components. They are at the origin of personal computer phenomena. Market of these kinds of computer is now dominant in computer industry. Production volume of personal computers is far above the best series of mainframe or UNIX workstations. It is now a mass industry of commonality products.

Around 1985, the evolution of VLSI processor has moved from a market pushed by technological evolution to a market pulled by requirements of end-users. The users are always willing to use newest software packages. To do that, they have to buy more powerful (personal) computers to run new software. Size of this market is so high that a large amount of money is required for technological improvements and the large demand for these computers makes possible a mass production of microprocessors.

As any other mass-production VLSI devices, price of microprocessors follows a classic curve starting from an introductory price around $1000 and decreasing in few years to $100. It is always difficult to imagine that price of such new extraordinary devices will fall down soon to the price of commonality goods.

4. LOOKING TO THE FUTURE

It is possible to predict computing power and size of future VLSI processors with a good accuracy. However, prediction of its architecture principle is more difficult to do, because the part of this curve representing the future also represents a challenge to designers of microprocessors.

We can extend up to 2010 the estimations on technology, complexity and perfor-mance of microprocessors. The results are very impressive. At this time minimum feature size of technology will be around 22nm, complexity of microprocessors around 1Gtransistors. Their computing power will reach 100Gips (giga instructions per second), taking in account their embedded capacity of multiprocessing.

Such predictions can be considered optimistic. However, past experiences have shown that linear prediction is a good predicting solution for this domain.

5. VLSI PROCESSORS ARE PRESENT IN ALL THE RANGE OF COMPUTERS

Computers built with microprocessors were progressively extended to all range of computers. They have already been present in a full range of minicomputer during the 1980s and in mainframes and UNIX workstations during 1990s. Around 1993, microprocessors have taken over mainframes in terms of complexity and perfor-mance. Now they are part of supercomputers since the first decade of 21st century.

Necessary large volume of microprocessors (more than a million units produced per year to be economically viable) leads to reduce the number of different models. As example, the volume of VLSI versions of mainframe or UNIX computers are not so large to economically compete with microprocessors designed for personal computers.

6.   MICROPROCESSORS ARE CHANGING OUR ENVIRONMENT

Microprocessors are increasingly playing a major role in modern society. They can be sorted in two classes: "visible" ones that are used to build different classes of computers, mostly personal ones, and "invisible" (embedded) ones that are used for controlling and monitoring machine tools, cars, aircrafts, consumer electronics and many other electronic equipments. The class of embedded microprocessors is the most important one in terms of the number of microprocessors in use. They are gradually changing relationship we have with these devices.

We can ask where this phenomenon is leading us. Microprocessors are changing insidiously our environment by making intelligent many objects that were previously simple ones. It is now possible to put a drop of intelligence into many devices with a little cost. We are only at the beginning of this deep societal evolution.

7.   THE COMPUTER HERITAGE

During their extraordinary evolution, microprocessors are using many technical innovations that have been developed for previous generations of computer, such as: pipeline architecture, cache, super-scalar architecture, RISC, branch prediction, etc.... All these features are used today in modern microprocessors that are probably the most advanced processors available on the market. They become followers of the computer story. For keeping the rate of their evolution new acceleration techniques have been developed.

Microprocessors become central components of computer designs. Building a processor with MSI technology is now completely obsolete and far outside economical optimum. Now, all new computers are based on the use of microprocessors (e.g. CRAY T3E supercomputer is constituted by an assembly of many ALPHA microprocessors).

8.   RISC AND CISC PROCESSORS

Classical processors are called CISC (for Complex Instruction Set Computers). They directly derivate from processors designed during 1960s and 1970s. Their main characteristics are:
− A (very) large instruction set using several instructions and data formats.
− Operands of instructions are fetched from registers and memory.

In 1975 John Cocke invents the notion of Reduced Instruction Set Computer (RISC) at an IBM laboratory. The use of the instructions by a CISC computer is

very unbalanced. For instance, the rate of use of the branch instructions is about
25% whereas the use of instructions manipulating a character string is far under
1%. By selecting a sub-class of instructions that is more frequently used, RISC
computers are well balanced and their hardware can be optimized. Several features
are specific to RISC processors:
- A large number of registers are used to decrease memory accesses.
- Few instruction formats are used to simplify instruction decoding.
- All instructions (except for memory accesses) are executed using a same number
  of clock cycles, simplifying sequencing of their execution.
- Memory accesses are only done by few specific instructions (that perform move
  operations between memory and registers)
    Hardware / software interface of RISC processors is lower than in CISC proces-
sors. This low interface comes from reducing the number of functions performed by
hardware and from increasing complexity of those executed by software. Programs
written for RISC processors are larger than those written for CISC ones. But, the
simplicity of RISC instructions allows a very good optimization of programs by
compilers. This optimization and short clock cycle of RISC computers give them a
higher global performance than CISC ones.
    Hardware of RISC microprocessors is far simpler than those of a CISC one. Most
popular series of old microprocessors were CISC ones (as the x86) but nowadays
most of microprocessors are RISC ones.

9.   MICROPROCESSORS AS VLSI CIRCUITS

9.1   Technologies and Architecture

Main characteristics of VLSI technology that are leading to overall organization of
microprocessors are:
- The cost of elementary cells of computing (e. g. a two bits basic operator needs
  around 35 transistors) that is higher than the cost of memory cells (two to six
  transistors).
- The speed of computing elements (few picoseconds) that is far faster as the
  practical need for computing results (few tenth of second)
  These differences lead to:
- Concentrating memory elements in large pure memories without any computing
  functions.
- Reusing as much as possible of the computing elements for many different
  operations in a serial way.
    These two characteristics lead basic structure of any processor becoming a loop
(Figure 4) where few computing elements fetch data from memories where they
store their results. Behavior of this loop is controlled by instructions.
    Another general characteristic of technology is the fact that speed of memory
elements is inversely proportional to their size. To obtain a "virtual" large and fast
memory, memory elements are organized in a hierarchy (Figure 5) where pertinent

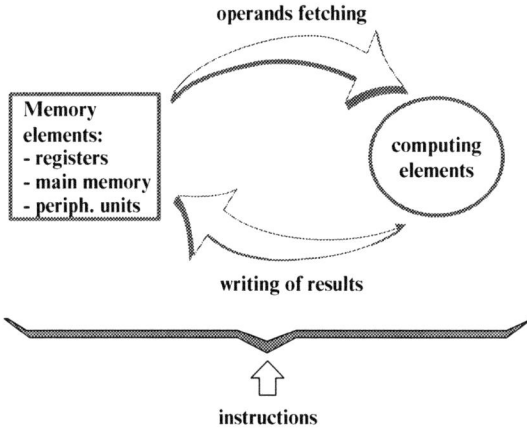

Figure 4. Basic computing loop

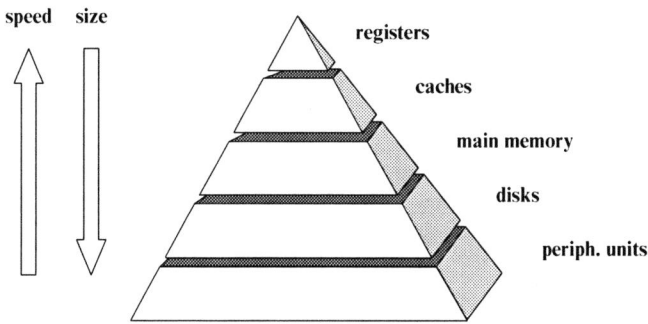

Figure 5. Memory hierarchy

data moves to be used from fast and small memories whereas data modifications comeback to update large and slow memories.

9.2   The Internal World of a VLSI Circuit

Microprocessors are obviously VLSI circuits. Each of these chips holds an internal world that is far smaller and faster than external one: capacitance are measured in ff, current in μa, size in μm, delay in ps. Internal gates of such an integrated circuit are running far faster than those of the Medium Size of Integration (MSI) level.

Passing a signal from the internal world of a VLSI chip to the external world is very costly. Multi-level amplifiers and geometrical adapters, provided by the packaging are necessary to perform this translation. Difference between internal and external worlds is comparable to driving electromechanical devices from an electronic board.

This cost leads VLSI designers to put as much as possible functional blocks inside a single chip instead of using multi-chip architectures multiplying (costly) interfaces. This feature leads to increase integration level and complexity of chips.

Another important issue is the distance between components inside these chips. At its scale a chip is a very large world! We can compare it to a country 2,300 km wide (as large as Europe) with roads 10m wide. Such a large area must be organized using geographical principles.

Cost of transferring information from one corner of a circuit to the opposite one is high. Powerful amplifiers are necessary. The relationship between different blocks must be carefully studied to increase local exchange and to decrease as much as possible long distances and external communication.

10.   PROCESSORS SEEN AS INTERPRETORS

A processor can be seen as an interpreter of its Instruction Set Architecture (ISA) [ANC 86]. The interpretation algorithm is executed by the hardware and eventually also by a microprogram. Many tricks are used to speed-up execution of this algorithm, mostly by introducing parallelism in its execution.

Functional organization of a processor is basically split into two main blocks called: control-section and datapath. The specification of these two blocks is defined by the decomposition of an interpretation algorithm into its control structure and its set of operations. The control structure of this algorithm is used to specify the control-section and its set of operations is used to specify datapath.

Since the mid 1970s, these two functional blocks are implemented as two separated pieces of layout that are designed with specific techniques.

10.1   Datapath Design

Datapaths are composed by operators, buses and registers. Very powerful techniques are used to design their layout. A basic technique to design data-path consists of using bit-slice layout strategy. A rectangular layout is obtained (Figure 6) by juxtaposition of layout slices. Each of them represents a single bit of data for the whole datapath. Assembly of cells by simple abutment constitutes these layout strips. These cells must have a fixed height. Buses are included over the layout of each cell.

The resulting layout is very dense and very optimized. Datapath is probably the best layout piece of a microprocessor. Its area is between a quarter and half of the whole area of a microprocessor itself (without including cache memory and other blocks included in a microprocessor chip).

10.2   Control-Section Design

Unfortunately, there is no global technique to design control-sections. It is not really a single block of layout but the assembly of several blocks like ROM, PLA, decoders and random logic. Layout of control-section is designed by traditional techniques of placement and routing.

control lines and test

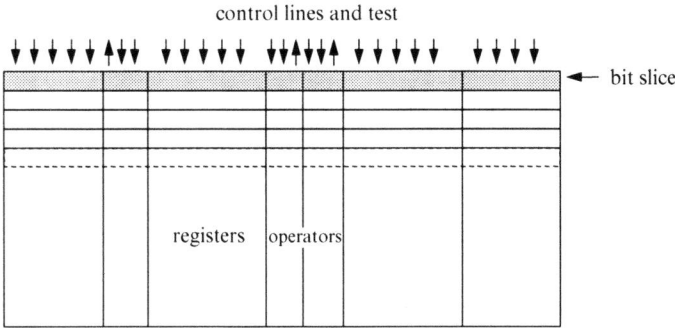

Figure 6. Layout organization of a data-path

The are several architectural approaches to design control-sections. Some of them come from traditional computers. But, opportunity to design specific blocks at VLSI level opens to new interesting techniques.

The control-section of a microprocessor can be microprogrammed or hardwired. A microprogrammed control-section (Figure 7) is designed around a ROM where microinstructions are stored. A microprogram address counter is used to address this ROM. Microinstructions are read from a ROM into a microinstruction-register where the different fields are decoded. Commands for the datapath are obtained by validating the outputs of this decoding with timing signal provided by the clock circuitry. Microprogrammed control-sections are more adapted to execution of complex instructions. These are executed by a variable number of clock cycles. This technique is mostly used by CISC microprocessors.

A hardwired control-section (Figure 8) can be designed around a PLA (Programmable Logic Array) performing instruction decoding. Specific gates are used to validate results of this decoding using timing signals provided by the clock circuitry, in order to obtain the commands for datapath. Hardwired control-sections are

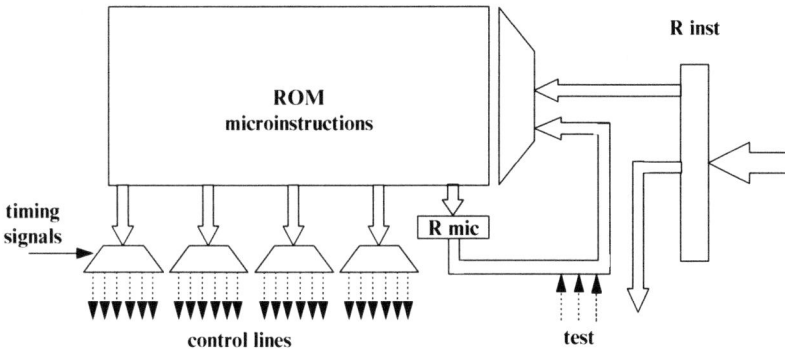

Figure 7. Microprogrammed control-section

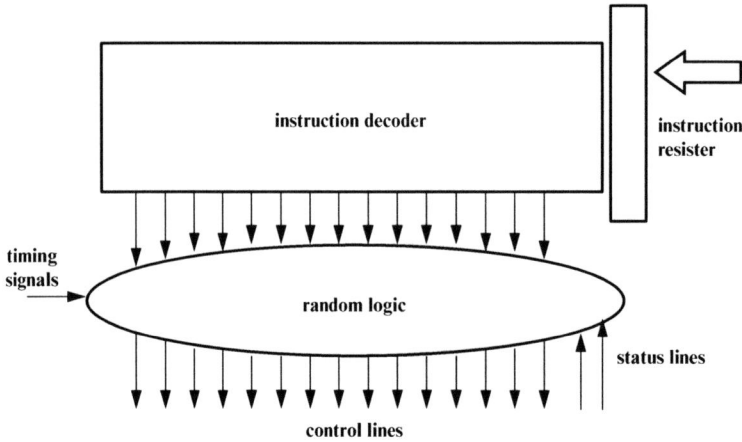

Figure 8. Hardwired control-section

more adapted to execution of simple instructions that uses a fixed timing pattern.
They are used in RISC microprocessors.

Both types of control-sections are composed by random logic blocks automatically designed using CAD tools.

11.   CODE TRANSLATION

The variable format of CISC instructions is incompatible with hardware constraints
of the fast processors. In order to keep the value of software investments and to
benefit of performances of new architectures, CISC instructions must be translated
to RISC ones (or into an other format suitable for fast execution).

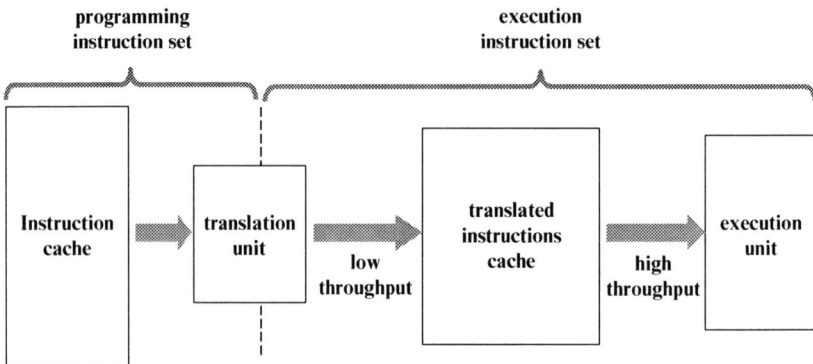

Figure 9. Use of a translation unit

This translation can be done either by:
- Specific hardware.
- Software executed by the processor itself when it needs new instructions to execute.

A hardware translator can be a simple look-up table performing translation into one, or a sequence, of executable instructions.

Using such a translator makes execution mechanism independent of ISA instruction set.

A cache can be introduced between translator and execution units (Figure 9). Its purpose is to divide the execution throughput for the translator. But the drawback is that translated code must be re-usable.

12.   SPEEDING-UP EXECUTION

The pressure to increase the performance of microprocessors leads their designers to find new ways to design these machines. Most of used acceleration techniques are based on superposition at the execution of the interpretation algorithm. The execution of new instructions starts before previous ones termination.

12.1   Pipeline Execution

The base of this execution technique is to slice the execution of an instruction into several sub-steps that are executed by several hardware blocks working like an "assembly-line" called Pipeline (Figure 10). Each stage of the pipeline performs a sub-step of execution. It gets data (instructions to be executed) from input registers and puts the results in its output registers. All these registers constitute an execution queue where instructions are progressively executed. A pipeline processor is simpler to design when all instructions have a same format (as in RISC). Pipeline processors executing CISC instructions must have a translator unit. Hardware complexity of a pipeline computer is larger than those necessary for sequential execution.

At each clock cycle, a new instruction starts execution by entering in the pipeline (Figure 11).

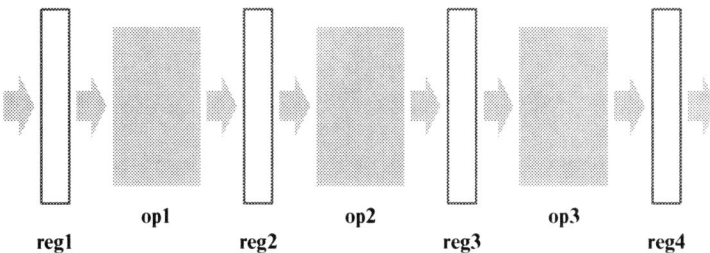

Figure 10. Hardware structure of a pipeline computer

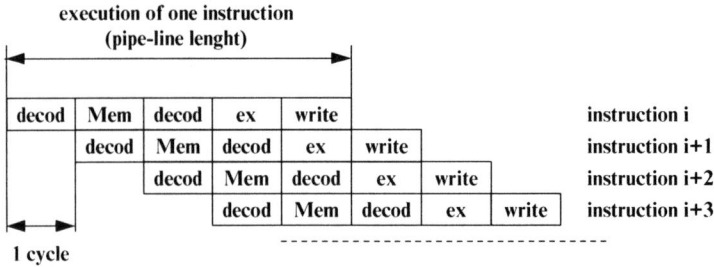

Figure 11. Instructions steam in a pipeline

The main drawback of the pipeline architecture is the dependency problem. The are two main cases of dependency:
– The first case is called data dependency. It occurs when an instruction in the pipeline wants to fetch data that has not yet been computed. This missing data must be computed by another instruction deeper in the pipe.
– The second case of dependency is called instruction dependency. It occurs when the computer does not know where is the next instruction to be fetched, because the condition used in a conditional branch is not yet evaluated.

Obviously, dependencies can be solved by stopping problematic instructions (and the following ones!) until missing data or missing condition is computed. The cost of this technique is very high because statistical measurements show that the probability of such a dependency becomes high after 3 instructions.

12.2   Branch Prediction

Making hypothesis on the behavior of conditional branches can reduce instruction dependency. Better the hypotheses are when better is performance improvement. When such a hypothesis becomes false, the instructions fetched in the pipe must be dropped. To allow such a backtracking, the instructions fetched upon a hypothesis are marked as conditional. These instructions are not allowed to modify ISA context of the computer (i.e. writing into programming registers and into memory).

Several techniques are used to predict branch instruction behavior. We will only discuss the dynamic ones.

12.2.1   Dynamic prediction techniques

Dynamic branch prediction techniques are based on the measurement of behavior of each branch instruction:
– The first technique developed by James Smith in 1981 consisted in recording the behavior of each branch instruction in order to predict the current one. Obviously, the buffer size used is limited and only the behavior of last branch instructions can be recorded. Probability of success of this technique is around 80%. This technique is used in ALPHA and K5 microprocessors.

- An evolution of the last technique called BHT (for Branch History Table) (Figure 12) use a saturating counter as an inertial element to measure behavior of each branch instruction. The probability of success goes up to 85%. This technique is used in many microprocessors like: PPC 604-620, ALPHA, Nx586, Ultra Sparc, M1, Pentium.
- Yeh and Patt [YEH 91] have proposed a two level dynamical mechanism in 1991. This technique consists in recording successive behaviors (behavior profile) of each branch instruction. In addition, a saturating counter is used to predict future behavior for each recorded profile. Probability of success becomes very high (95%). This technique is used in Pentium-Pro microprocessor and its successors.

12.3   Super-Scalar Execution

Another acceleration technique consists in setting several pipelines in parallel in order to increase the throughput of processors. These pipelines can be specialized or identical. Some processors use a specific pipeline for integer computation and another for floating point. In this case each pipeline fetches only instructions that it is able to execute. Specialized pipelines use different set of resources (registers). They is no dependency between them. When pipelines are identical they share many resources (registers) and many conflicts and dependencies can occur. As an example, ALPHA microprocessor uses two identical pipelines.

The term "superscalar" have been extended from this original definition to the capture of a large set of techniques used to speed-up processors.

12.4   Data-Flow Processors

The most powerful technique to speed-up execution is a data-flow execution technique [GWE 95a] also called desynchronised super-scalar. This technique comes from the HPS (High Performance Substrate) project developed in the University of Berkeley in the 1980s [PAT 85a-b].

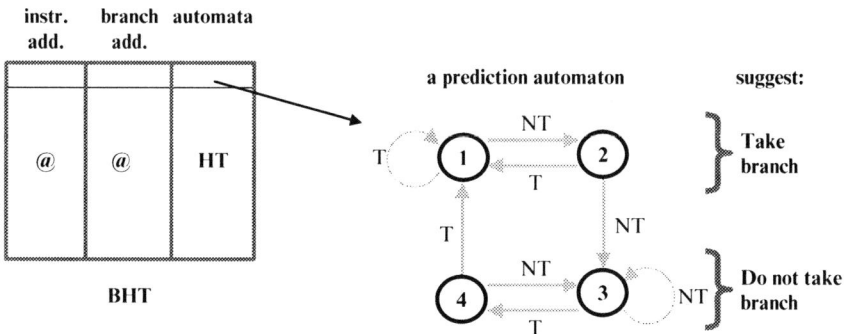

Figure 12. Branch History Table

Data-flow execution technique consists of executing each instruction as soon as possible without waiting for the completion of previous ones (except for dependencies). All functions participating on the execution are organized as independent agents acting over an instruction when their necessary information (or data) is present (e.g. agent that load data from memory to execute a load instruction when its address is present). This kind of working organization is dual of an assembly line and can be called "assembly hall" where objects to build (e. g. aircrafts) stay in place and workers move to objects that are waiting for their services.

Intel Pentium Pro, II – 4 [COL 06] and AMD K6 – 8 [SHI 98], Athlon are using this execution technique.

12.4.1   Register renaming

Register renaming is the main step in the translation from a classical instruction set to a data-flow one. Its purpose is to translate the use of classical reusable ISA registers into the use of single-assignment ones. This translation breaks the different use of each ISA register into the separated use of several different physical registers.

ISA registers are mapped into a memory that is several times larger than the number of ISA registers. Current status of this mapping is kept in a physical table called RAT (for Register Allocation Table) (Figure 13). Each entry of RAT corresponds to an ISA register and contains the address of a physical register containing its current value. Fields naming ISA (operand) registers in an instruction are replaced by the addresses of physical registers mapping these ISA registers. To make physical registers single assignment ones, an entry of RAT (corresponding

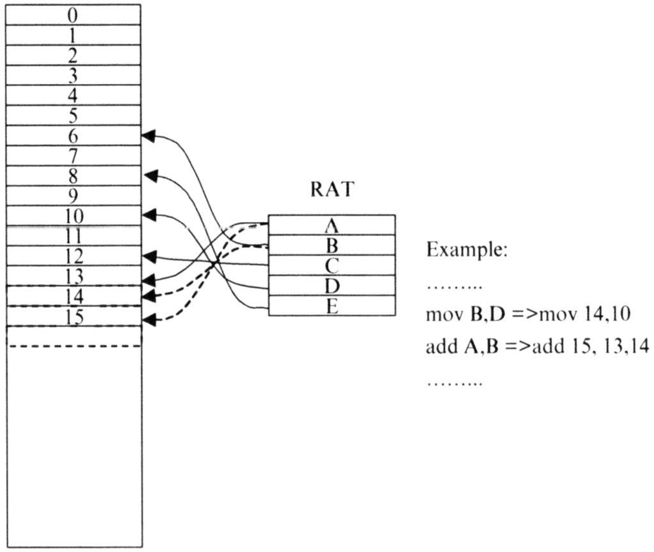

Figure 13. Register renaming

to an ISA register) is changed to designate a new free physical register for each load of this ISA register. Field naming ISA destination register of an instruction is replaced by the address of this new physical register.

Saving the contents of the RAT allows a very fast backtracking when a conditional instruction stream must be aborted. Duplicating the RAT allows to start multiple threads running in parallel.

Renaming of registers eliminates pseudo-dependencies corresponding to access to reusable ISA registers seen as single resources. For this reason, this technique can also be used in pipeline and super-scalar architectures.

12.4.2   Data-flow execution

The first step in a data-flow execution is register renaming that transforms ISA instructions into data-flow ones. The second step consists in putting instructions into a large buffer called ROB (for ReOrdering Buffer) that is an assembly hall. In this buffer, flags are added to instructions to indicate when their operand registers (renamed) are filed.

The basic execution loop of data-flow computers (Figure 14) is organized as a set of agents getting from the ROB instructions that are ready to be executed. When getting their instructions, these agents fetch values of their operands from registers and execute them. Their results are stored into destination registers. To close the data-flow loop, associative searches are performed into ROB indicating to others instructions that these registers are filled (these operands become ready).

Pure data-flow execution is independent of the order of instructions in ROB because renamed instructions are fully independent. However, the use of a limited

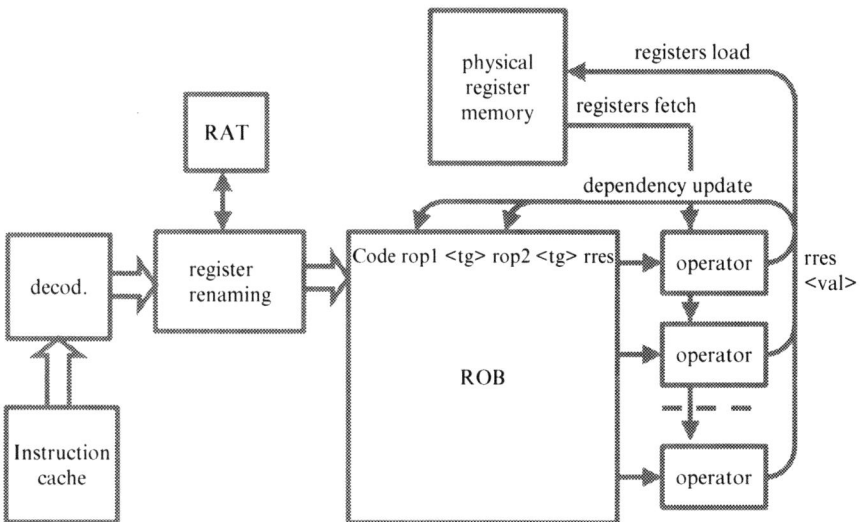

Figure 14. The data-flow loop

number of resources (physical registers and main memory) re-introduce serialization problems requesting that instructions must be inserted and extracted from ROB in an execution order:
– The memory used to hold physical registers has a limited size. It is necessary to free a physical register position when it is no longer used. Finishing an instruction that store an ISA register indicates that the previous physical register used to hold this ISA register is free.
– In order to keep main memory consistency, its accesses must be done in the execution order. For that, memory stores are put into a FIFO queue. In normal mode queued stores are executed when memory is available. In the conditional mode, writing into memory becomes forbidden until the hypothesis is not yet confirmed. In every case, reading of a value in conditional mode is possible by starting an associative search into the queue before accessing the memory.

12.5   Very Long Instruction Words (VLIW)

In VLIW computers, several instructions (having no dependencies) are packed together in long words ended by branch instructions. These packed instructions are executed in parallel by several hardware pipelines (Figure 15). Compilers have to prepare these long words for this kind of computers.

At each clock pulse, a long word is extracted from cache to feed the execution pipelines running in parallel. TMS 320 and Intel Itanium® processor are using this execution technique. Transmeta CRUSOE also uses this technique to execute PC code by translating it by software during its wait states.

12.6   Multimedia Facilities

Modern processors execute specific SIMD (Single Instruction, Multiple Data) instructions for multimedia processing (video, sound,…). Large registers (128 bits) are loaded with sub-strings of multimedia data. Specific instructions perform the

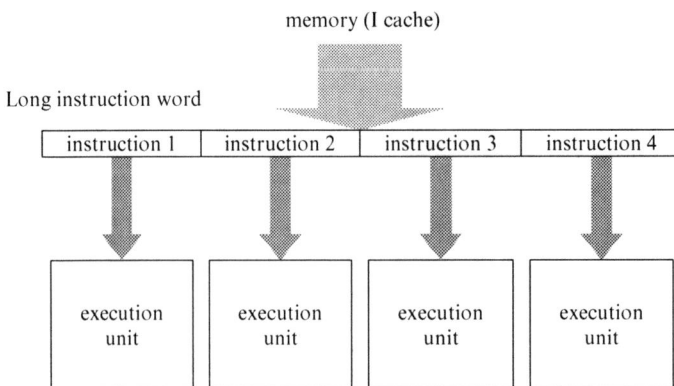

memory (I cache)

Long instruction word

| instruction 1 | instruction 2 | instruction 3 | instruction 4 |

| execution unit | execution unit | execution unit | execution unit |

Figure 15. VLIW processor

same operations on all data of such large registers. Some operations are saturated, this means that an overflow to a maximum value keeps this maximum value instead switching to the minimum one by a modulo effect.

12.7   Multiprocessor in a Chip

Data-flow processors can be easily organized to execute several programs (multi-threading) in pseudo parallelism, but the overall performance stays the same of a single processor.

For long-time, multiprocessor systems was a desire of computer scientists looking for getting more and more computing power. The rise of power dissipation toward extreme values has leaded main manufacturers to develop monolithic multiprocessors, called multi-cores. Such an idea is not new and several projects have already been investigated (As example a project of double MC 6802 had been investigated at the end of 1970s). Today, the number of elementary processors in a multi-core chip is small (2 to 8) but the evolution of such a concept can be pushed toward systems composed by a big set of cores in a chip.

13.   CONCLUSIONS

Microprocessor evolution is one of the most exciting technical stories of humanity. Computing power of these devices has been multiplied by a factor of several thousand during last 35 years. No other technical domains have had such an evolution.

Microprocessors are progressively substituting themselves to other technologies of processors. Probably they will be the component of all range of supercomputers at the beginning of this century.

Nowadays, processors can be seen as a set of several sub-processors working together to execute a single program. Microprocessor chips are now holding several processors working in parallel on several program threads.

We can ask where such an evolution leads us. Estimated future computing power is very impressive and will be available to anyone to use the best techniques for simulation, CAD, virtual worlds, etc... making personal computers the best tools we have ever had.

REFERENCES

[TOM 67] R. M. Tomasulo, An Efficient Algorithm for Exploiting Multiple Arithmetic Units, IBM Journal of Research and Development, Vol 11(1), Jan. 1967, p25-33
[PAT 85a] Y. N. Patt, W. W. Hwu, M. C. Shebanow, HPS, a New Microarchitecture: Rationale and Introduction, in Proceeding of the 18th International Microprogramming Workshop, Asilomar, Dec. 1985.
[PAT 85b] Y. N. Patt, S. W. Melvin, W. W. Hwu, M. C. Shebanow, HPS, Critical Issues Regarding HPS, a High Performance Microarchitecture, in Proceeding of the 18th International Microprogramming Workshop, Asilomar, Dec. 1985.
[ANC 86] F. Anceau, The Architecture of Microprocessors, Addison-Wesley, 1986

[TRE 86] N. Tredennick, Microprocessor Logic Design, Digital Press, 1986
[HEN 90] J.L. Hennessy and D.A. Patterson , Computer Architecture a Quantitative Approach, Morgan
    Kaufmann, 1990
[YEH 91] T-Y. Yeh and Y. N. Patt, Two-Level Adaptative Trainning Branch Prediction, ACM/IEEE
    24th Ann. Int'l Symp. on Microarchitecture, Nov. 1991, p51-61
[COM 92] R. Comerford , How DEC developed Alpha, IEEE Spectrum, July 1992
[GWE 95a] L. Gwennap, Intel's P6 Uses Decoupled Superscalar Design, Microprocessor Report, Vol 9
    n 2, Feb 16, 1995
[GWE 95b] L. Gwennap, The Death of the Superprocessor, Microprocessor Report, Vol 9 n 13, Oct 2,
    1995
[GWE 96] L. Gwennap, Intel's MMX Speeds Multimedia, Microprocessor Report, Vol 10 n 3, March
    5, 1996
[MIL 96] V. Milutinovic , Surviving the Design of a 200 Mhz RISC Microprocessor, Lessons Learned,
    Computer Society, 1996
[GWE 97] L. Gwennap, Intel, HP Makes EPIC Disclosure, Microprocessor Report, Vol 11 n 14, Oct
    27, 1997
[SHI 98] B. Shiver and B. Smith, The Anatomy of a High-Performance Microprocessor, A System
    Perspective, IEEE Computer Society, 1998
[COL 06] R. P. Colwell, The Pentium Chronicles, Wiley-Interscience, 2006

BIOGRAPHY

Prof. Anceau received the *"Ingénieur"* degree from the *Institut National Polytech-nique de Grenoble* (INPG) in 1967, and the *Docteur d'Etat* degree in 1974 from the University of Grenoble. He started his research activity in 1967 as member of the *Comite National pour la Recherche Scientifique* (CNRS) in Grenoble. He became Professor at INPG in 1974 where he led a research team on microprocessor archi-tecture and VLSI design. In 1984, he moved to industry (BULL company, close to Paris) where he was lead a research team on Formal Verification for Hardware and Software. The first industrial tool for hardware formal verification and the technique of symbolic state traversal for finite state machines was developed in this group. In 1996 he took his present position as Professor at *Conservatoire National des Arts et Metiers* (CNAM) in Paris. Since 1991 he has also been a Professor at *Ecole Polytechnique* in Palaiseau, France. Since October 2004 he is doing research activ-ity in the ASIM/lip6 laboratory in the University *Pierre et Marie Curie* in Paris. His main domains of interest are: microprocessor architecture and VLSI design. He has given many talks on these subjects. He his the author of many conference papers and of a book entitled *"The Architecture of Microprocessors"* published by Addison-Wesley in 1986. He launched the French Multi-Chip-Project, called CMP, in 1981.

CHAPTER 5

PHYSICAL DESIGN AUTOMATION

RICARDO REIS[1], JOSÉ LUÍS GÜNTZEL[2], AND MARCELO JOHANN[1]

[1] *UFRGS – Instituto de Informática- Av. Bento Gonçalves, 9500 Caixa Postal 15064 – CEP 91501-970, Porto Alegre, Brazil, e-mail: reis@inf.ufrgs.br, johann@inf.ufrgs.br*
[2] *UFPel – Departamento de Informática, Caixa Postal 354 – CEP 96010-900, Pelotas, Brazil e-mail: guntzel@ufpel.edu.br*

Abstract: The paper addresses physical design automation methodologies and tools in which cells are generated on the fly as a form of contributing to power and time optimizations and improving the convergence of the design process. By not being restricted to cell libraries it is possible to implement any logic function defined at logic synthesis, including static CMOS complex gates – SCCG. The use of SCCG reduces the amount of transistors and helps to reduce wire length and static power. Main strategies for automatic layout generation, like transistor topology, contact and via management, body ties placement, power lines disposition and transistor sizing come into account. For both standard cell and automatic layout generation methodologies, a good set of placement and routing algorithms is needed. This work addresses convergence issues of the methodology as well, including some key strategies that help the development of efficient CAD tools that can find better layout solutions than those from traditional standard cell and fixed-die methodologies

Keywords: Automatic Layout Synthesis, Physical Design, EDA, Microelectronics

1.   INTRODUCTION

Recent fabrication technologies are changing some physical design paradigms. Many electrical effects that were neglected in the past are getting increased importance and cannot be neglected anymore. The design of random logic blocks is still based on the use of cell libraries. This was a good solution for a long time because the timing of a circuit could be calculated by considering only the gate delays. The standard cells approach consists on assembling a block layout by instantiating the layouts of cells from a library. Generally, a standard cell library provides few layout versions for each logic gate type, each of them sized to drive different fan-out charges. As long as the cells were previously characterized in terms of delay and

R. Reis et al. (eds.), Design of Systems on a Chip, 83–108.
© 2006 *Springer.*

power, it was possible to estimate timing and power of an assembled layout by using only cells parameters and fan-out information.

Nowadays, as clearly stated in the Semiconductor Roadmap [1], the connections are taking the main role in the timing calculation. Thus, the pre-characterization of cells does not lead anymore to valid performance estimates because it is necessary to take into account the effects of routing wires as, for example, the capacitive coupling. As a consequence, the predictability claimed as a main advantage of the standard cell methodology no longer holds. Hence, one of the current challenges in EDA is to develop tools able to accurately estimate the effects associated to the interconnections. Another point to consider is that the cells in a library do not have flexibility relating to transistor sizing. A cell library normally has cells designed for area, timing or power, but only one cell for each objective. As the cell library is designed to support a large fan-out, cells are normally oversized with respect to the size they should have, when considering the environment where they are inserted. Indeed, if many cells are larger than they should be, the final layout will be larger than it could be. In addition, the average wire length will be longer than needed. Currently, it is very important to reduce wire lengths because they have a great influence on circuit performance. Wire length reduction may be achieved by using automatic layout generation tools. Some designers have claimed that it is difficult to estimate the performance of circuits generated by such techniques because cells were not pre-characterized. However, in deep submicron CMOS technologies it is not possible to estimate performance based only on pre-characterization information because the influence of the routing became dominant, and also due to physical effects of fabrication that make layout structures sensitive to their surrounding environments. This issue renders performance estimation equally difficult for both standard cell and automatic generation approaches. To overcome this problem, accurate power and timing estimation tools able to consider the interconnection contribution are needed. But with the synthesis of cells on the fly, during layout synthesis, it is possible to reduce the number of transistors and to reduce the average wire length. It also allows appropriate transistor sizing, and consequently to increase performance. This paper will give more details about physical design automation and how it can improve layout optimization. The paper also states that the challenge we have nowadays is the development of efficient tools for layout design automation where the cells are designed to fit well in their specific environment. This fact has already caught the attention of many researchers and CAD vendors. Traditional companies as IBM have already developed custom module generation tools in the past, and new companies as Zenasis are beginning to sell tools for cell synthesis that aim to substitute standard cells in critical parts of a circuit [2].

2.   THE STANDARD CELL APPROACH

The standard cell approach is still accepted as a good solution for the layout synthesis problem. Due to the nature of the standard cell approach, it presents several limitations that were exploited to allow its automation. For instance, the technology mapping is limited to the cells found in a library and to the sizing possibilities also available in the library, which are restricted to few options for each logic gate.

The standard cell approach is currently the most used approach for the layout synthesis of random logic blocks. The main claimed reasons for that are the pre-characterization of cells (arranged as a library) and the lack of efficient cell design automation tools. The cell pre-characterization allows the designer to evaluate the delays of a circuit with a reasonable accuracy considering old fabrication processes, as well as a good estimation of power consumption. In the design of circuits using current submicron technologies, cell pre-characterization does not give sufficient information to evaluate the delays with the necessary accuracy because the delay related to the connections became prominent. The challenge is to have efficient physical design tools to perform an automatic full custom physical design, with cells automatically designed to meet the requirements of the environment where they will be inserted. These tools should include automatic circuit pre-characterization.

The increasing development in IC technology brought the transistor gates to submicron dimensions. These advances brought also the need to review layout design methods. Timing and power reduction are important issues to be considered at all design levels including layout design. Hence, the goal is to have a set of layout synthesis tools with the following features:

– **Specialized cells**. Each cell should be generated on the fly, considering the specific needs of the circuit environment where it is going to be placed, like cell fan-out.
– **Transistor minimization**. Reduction of the amount of transistors by the use of static CMOS complex gates (SCCG). The advantages of using SCCG gates are: area, delay and power reduction (when compared to the standard-cell approach).
– **Fan-out management**. The management of the fan-out should be used for tuning delay, power and routing congestion and for wire length reduction.
– **Transistor sizing**. Each transistor should be sized according to a good compromise between power and delay, considering the environment where it will be placed, for instance, its effect as driver/load in the signal path.
– **Placement for routing**. Placement algorithms should be guided to reduce the average wire length, routing congestion, and improve timing.
– **Full over the cell routing**. The routing should be done over the cells avoiding the use of routing channels and spaces.
– **Routing minimization**. Addressing interconnections routing to reduce the average wire length and to minimize the length of the connections on the critical paths and critical net sinks.
– **Performance estimation**. The delay of a circuit should be evaluated with a good accuracy before its actual generation. So, it is fundamental to have timing and power estimation tools that take care of both transistor's and interconnection's influence in circuit performance.

- **Area evaluation**. The area and shape of a circuit should be evaluated before the generation of each functional block. This is important to optimize the floor plan of an entire chip.
- **Technology independence**. All design rules are input parameters for the physical design tool set. The redesign of the layout of a same circuit with a different set of design rules should be done by just changing the technology parameters description file and by running the layout synthesis tool again.

3.   USING STATIC CMOS COMPLEX GATES

A very important advantage of the on-the-fly method relies on the flexibility to generate the layout of any type of CMOS static cell. This is because appropriate generation strategy and algorithms may be used, as the linear matrix style, for example.

A static CMOS gate is composed of two distinct networks of transistors: a PMOS transistor network, connected between the power supply (Vdd) and the gate output, and a NMOS transistor network, connected between the ground (Gnd) and the gate output. The PMOS and the NMOS networks have equal number of transistors and are made up from serial/parallel only associations, with the PMOS-network being the dual association of the NMOS-network. Each input of a static CMOS gate is connected to the gates of an appropriate NMOS/PMOS pair of transistors. Hence, the number of transistors in one network (PMOS or NMOS) is equal to the number of gate inputs, while the total number of transistors of the static CMOS gate is simply twice its number of inputs.

For any combination of logic values applied to the inputs, a static CMOS gate always gives a path between the output to either Vdd or Gnd nodes. This full restoring feature gives static CMOS gates a very robust behaviour thus being the designers' first choice in terms of logic design style. This feature also makes static CMOS gates very attractive for developing automatic layout generation tools. Figure 1 shows examples of static CMOS gates.

Static CMOS gates may be classified according to the type of association of its PMOS and NMOS transistor networks. A static CMOS simple gate is a static CMOS gate where each network is either a parallel-only or serial-only association of transistors. For instance, if the NMOS-network is a serial-only association, then the PMOS-network is a parallel-only association and the gate performs the NAND logic function. Frequently, the term "static CMOS complex gate" is applied to refer to the general case of static CMOS gates. In static CMOS complex gates (or SCCGs, for shortly) the NMOS and PMOS networks may be mixed associations of transistors (serial/parallel), which still follow the duality property as any static CMOS gate.

The various subsets of SCCGs may be defined by specifying the number of serial/parallel transistors encountered in the NMOS/PMOS networks. In terms of technology mapping strategies, the set of static CMOS gates formed by no more than n (p) serial NMOS (PMOS) transistors may be referred to as a "virtual library" [5]

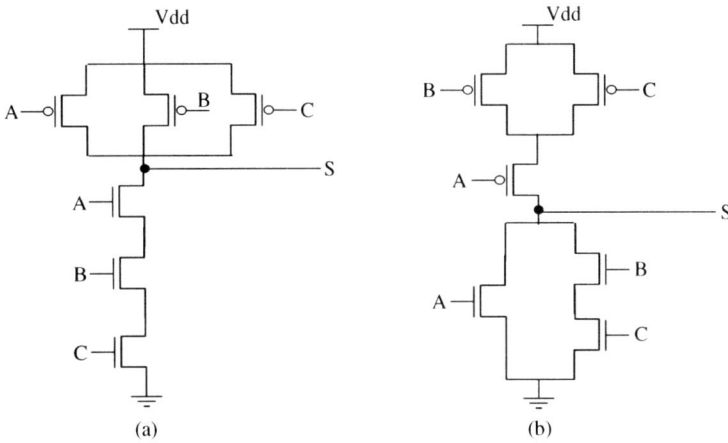

Figure 1. Examples of static CMOS gates: a 3-input NAND gate (a) and a static CMOS complex gate (b)

Table 1. Number of SCCGs with limited serial transistors, from [3]

Number of Serial NMOS Transistors	Number of Serial PMOS Transistors				
	1	2	3	4	5
1	1	2	3	4	5
2	2	7	18	42	90
3	3	18	87	396	1677
4	4	42	396	3503	28435
5	5	90	1677	28435	125803

and may be designated by SCCG (n,p). Table 1 (borrowed from [3]) shows the number of SCCGs for various virtual libraries, each library with a different value for the pair (n,p). Due to the electrical characteristics, it is not advisable to use SCCGs with more than 3 or 4 serial transistors in general. But in some cases a SCCGs with 5 or more serial transistors can provide a better performance when substituting a set of basic cells. Even when limiting the maximum number of serial transistors to 3, 87 possibilities of SCCGs could be used for circuit synthesis. When limiting to 4, the number of possible SCCGs grows to 3503. And limiting to 5 serial transistors, the set of functions goes to 28435. Such numbers of possibilities give the on-the-fly layout generation method a lot of flexibility.

The use of SCCGs may be explored within the technology-mapping step of logic synthesis in order to reduce the total amount of transistors used to implement a given logic. This is illustrated by the two mappings for the logic function $S = \bar{A}B + A\bar{B}$, as shown in Figure 2. Mapping this function with SCCGs results in 4 transistors less than mapping it by using static CMOS simple gates only.

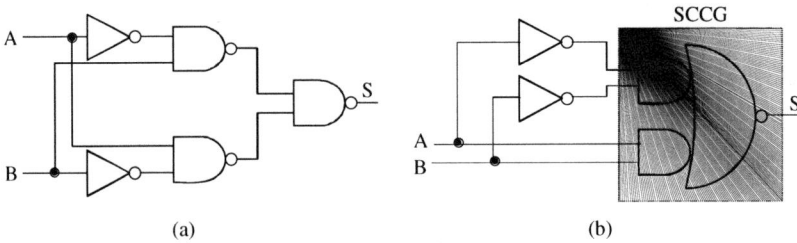

Figure 2. Mapping logic function $S = \bar{A}B + A\bar{B}$ using static CMOS simple gates only (a) and using SCCGs (b)

When using SCCGs with a maximum of 3 serial transistors, it is possible to reduce by 30% the number of transistors, when comparing to the same circuits mapped using static CMOS simple gates only, as shown in [4]. BDD-based technology mapping with SCCGs is presented in [5].

It is important to remark that by using fewer transistors, the resulting circuit area is also reduced. But if the circuit area is reduced, then the average wire length may also be reduced and thus, the critical delay may be smaller then that of the same circuit architecture, mapped to static CMOS simple gates. Furthermore, a circuit built up from fewer transistors tends to consume less power because both dynamic and leakage currents are smaller. It is well known that for current and upcoming submicron technologies, leakage current is a very hard issue that must be controlled.

4.   LAYOUT STRATEGIES

The layout synthesis tools should consider several goals as area reduction, wire length reduction, congestion reduction, routability, full over-the-cell routing and critical path management. The first challenge in automatic layout design is that the number of variables to handle is high and it is difficult to formalize rules to be used by a layout tool. When a designer is constructing a cell by hand, there are many choices for the placement of each cell rectangle. But man ability to visually manage topological solutions for a small set of components can generally find a better solution than that obtained by known tools. One can observe the high density that was found in random logic block circuits designed by late seventies (e.g. the MC 6800 and the Z8000 microprocessors). At that time all routing was performed over the cells by using only one metal layer and without using exclusive routing channels (Figure 3 shows an example). Even in present days no tool is able to achieve a similar layout compaction.

On the other hand, man cannot handle large circuits containing millions of rectangles without using CAD tools. In addition, the new fabrication technologies are demanding new layout generation solutions, able to treat the side electrical effects that were formerly neglected. Thus, there is a research room for the development of methodologies and tools that can efficiently handle all the possibilities and choices

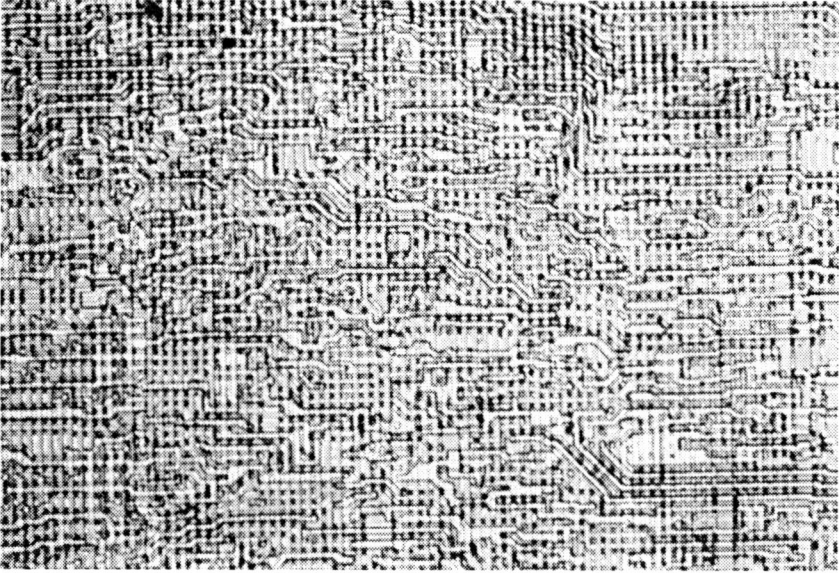

Figure 3. A random logic block detail taken from the Zilog Z8000, where the high layout compaction can be observed

in the topology definition of each cell. But to begin with, it is possible to set up some parameters and give flexibility to others. For example, the height of a strip of cells can be fixed and the cell width can change to accommodate cell's transistors.

The first step in the construction of a CMOS layout synthesis tool is to define the layout style, which corresponds to the definition of the following issues:
- allowed transistor topologies,
- power supply distribution,
- body tie placement,
- clock distribution,
- contact and via management,
- management of routing in polysilicon and metal layers.

4.1 Transistor Topology

Figure 4 shows some possible transistor topologies. The first decision to be taken while planning a layout generation tool regards transistor orientation. Two possible basic orientations exist: horizontal (Figure 4a) and vertical (Figure 4b). The horizontal orientation seems to be attractive because it allows all cells to have the same height, even when cells have different transistor length. However, the horizontal orientation increases the complexity of the intra-cell routing. Moreover, it renders more difficult to merge diffusion areas of neighbouring cells, thus resulting in a

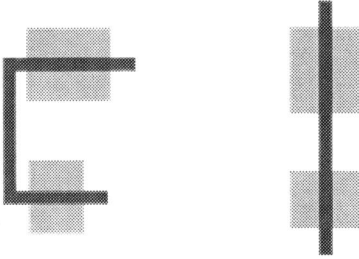

Figure 4. Possible transistor basic orientations: horizontal (a) and vertical (b)

waste of area and in greater perimeter capacitances. The vertical orientation, by its turn, facilitates the automatic input/output and intra-cell routing because contacts may be placed following a virtual grid. In addition, transistors of the same type that are laid out vertically may easily share their source/drain terminals, hence reducing the associated perimeter capacitances and also the amount of wasted area. Figure 5 illustrates the possibility of merging source/drain of neighbouring cells. It also highlights that by merging source/drain, the resulted diffusion area assume a "strip" shape (possibly with "gaps").

Due to the mentioned reasons most of the existing automatic random layout generators adopt the vertical orientation for the transistors. The vertical orientation presents some limitations, however. For instance, accommodating transistors with different lengths in a given diffusion strip results in waste of area, as shown in Figure 6a. A possible solution to overcome this problem relies on allowing transistor to be sized only in a discrete manner, with the length being an integer multiple of the standard length (the standard is defined as a generator input parameter). By doing so, it is possible to keep the height of a given row by using the folding

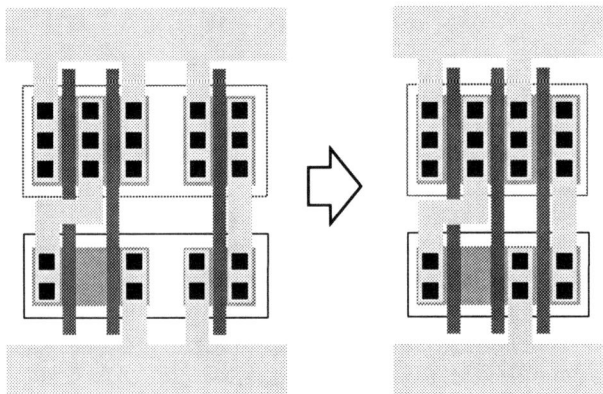

Figure 5. Merging source/drain areas of adjacent cells when the vertical orientation is used

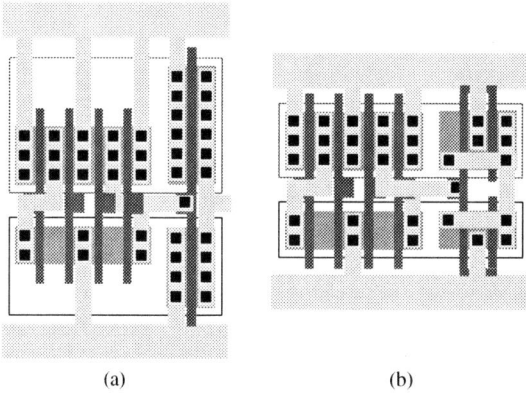

Figure 6. Waste of area resulting from the continuous sizing (a) and the area optimization resulting from the use of discrete sizing along with the folding technique (b)

technique. Using such technique a transistor that is to be sized to three times the standard length is laid out folded into three equal parts that are connected together. The folding technique is shown in Figure 6b. Although the folding technique (along with the discrete sizing) reduces the layout generation flexibility, it is a practical and feasible solution that leads to good results.

Another problem in the layout generation using vertical transistors concerns the misalignment between the gates of the PMOS-NMOS transistor pairs that are to be connected together. Analyzing the layouts shown in Figure 6 one may notice that it is possible to reduce the distance between adjacent gates when the diffusion area between such gates does not hold a contact. By doing so, a misalignment between the gates of a complementary PMOS-NMOS transistor pair appears (as shown in Figure 7). Therefore, doglegs must be inserted in the polysilicon that

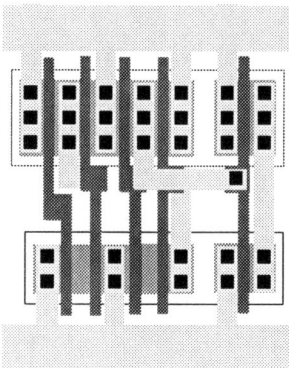

Figure 7. Misalignment between the gates of PMOS-NMOS transistor pairs

connects the gates of a PMOS-NMOS transistor pair. This may result in waste of area and worse electrical performance. Furthermore, for rows with many transistors the accumulated misalignment may be significant causing so many doglegs that the overall circuit electrical performance may be affected. Thus, other strategies must be used in order to reduce the misalignment of transistors. One possibility is to use some (or all) of the diffusion strip gaps to accommodate the contacts to the substrate (also called "body-ties"). Other possibility is simply to enlarge the strip gaps to compensate for the misalignment, using such spaces as a possible contact/via slot.

Besides vertical and horizontal transistors it is also possible to design transistors with doglegs in the polysilicon that defines the transistors. This can be interesting in technologies where it is possible to use only one contact between diffusion and the metal that handles the VDD or GND. It is sure that the inclusion of transistors with directions others than horizontal or vertical significantly increases the complexity and running times of the layout CAD tool.

4.2   Power Supply Distribution and Body Tie Placement

VCC and Ground are distributed in metal layer generally using an interleaved comb topology, as shown in Figure 8. Such topology is widely used because the voltage trop is such that at any point of the structure the difference of potential between VCC and Ground lines is the same.

Naturally, the final distribution of power supply must be in first or second metal layer, but upper metal layers may also be used to distribute power to the entire chip. In the example showed in Figure 8 cells are placed in the space between a VCC and an adjacent GND line, forming a row of cells. In order to avoid waste of area, two adjacent rows share a power supply line (either GND or VCC). Therefore, half of the cell rows are mirrored. Another feature in the example of Figure 8 that is

VCC

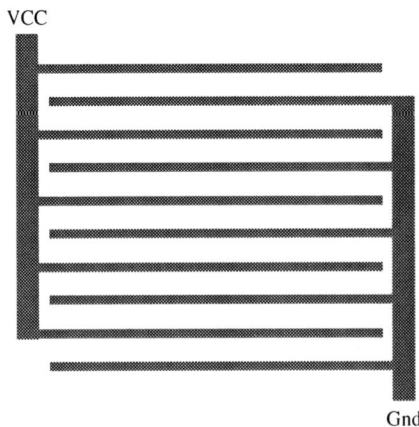

Gnd

Figure 8. Example of interleaved comb topology for the VCC and Ground distribution network

worth to mention is the power supply lines run outside the cell area. However, there are two other possible strategies: the power supply lines may run over the diffusion strip or may run in the middle of the cell rows, more exactly between the PMOS and the NMOS diffusion strips. Each of these two latter strategies presents important drawbacks. In the case of running power lines in the middle of cell rows, the region between PMOS and NMOS diffusion strips is higher and thus the polysilicon line connecting the gates of PMOS-NMOS transistor pairs is longer and hence are more resistive. Therefore the electrical performance of the cells is worse. In the case of power lines running over the diffusion strips, as long as the height of transistors is limited, the power lines may run in first metal layer to ease the contact to the underlying diffusion. However, the connections inside a cell also use the first metal layer. As a result, there is less flexibility to complete the intra cell connections. But power lines running over transistors can use metal 2. In this case the internal routing of cells is easier.

There are two main strategies to place the contacts to the substrate (body-ties): place them below the power supply lines or in the gaps of diffusion rows. The most practical one is placing the body-ties below the power distribution lines because in this case no extra connections are need. But the placement of body-ties in the gaps of diffusion may help to reduce the distance between power lines and the p and n diffusions.

4.3 Contact and Via Management

Contact holes, or simply contacts, correspond to the place where the lower most metal layer achieves the diffusion or the polysilicon underlying layer. In other words, contacts are the only way to physically achieve the circuit devices. Thus, any automatic layout generation tool must incorporate good strategies in placing contacts in order to guarantee full accessibility to the gates inputs/output.

Typically, the number of terminals to access the inputs of a cell ranges from one to four. In the cells of Figures 6 and 7, for example, it is only possible to place the input terminal in the middle of the cell, in a predefined place. An alternative strategy could be previewing other possible places for the input terminal. Then, several possibilities could be offered to the router. The number of possibilities can be reduced to improve tools running times.

Good strategies for placing vias are equally demanded for the success of the routing. Contacts and vias are distributed as real towers of vias within the circuits, when connecting distant layers. A wire in an intermediate layer should not prevent the construction of stacked vias to bring a signal to the upper layers.

4.4 Routing Management

Although current fabrication technologies provide the designer with many metal layers, a good routing management is still crucial because the number of transistors that may be integrated in the same chip is extremely high and new side effects are constantly appearing.

There are two fundamental rules that one encounter in any routing management strategy. The first one consists in using each metal layer to perform connection in only one direction, while adjacent layers being always in orthogonal directions. The only exception is the case where a signal changes from one track to the adjacent one, the so-called "dogleg". Since the two used tracks are adjacent, no other signal could pass between the two tracks, and therefore it is no worth to change to an upper metal layer.

The second strategy relies on reserving the lower metal layers to local connections, mainly intra cell connections, while the upper layers are used preferably for very long connections and global signals, as the clock. The intermediate layers may be used to perform inter block connections and busses.

Besides these naïve directions, other ones may contribute to the routing tool in completing all connections. For example, according to the row height a certain number of metal tracks may run over the cells. Considering the lower most metal layers, it is advisable to reserve some of those tracks to the intra cell connections and others to reach the input/output terminals of the cells. This greatly helps in avoiding terminal blockage.

Another routing strategy that is applicable to CMOS technologies with a small number of metal layers is allowing the tool to route some long connections in a channel. A routing channel is an obstacle free area (i.e., without transistors) that is dedicated to perform the routing. Since the routing channel decreases the transistor density, it should be avoided whenever possible. However, it is preferably to reduce transistor density than to fail in performing all connections. If the router is not able to complete a single connection, the layout cannot be sent to fabrication!

4.5   Putting Strategies Together

Figure 9 shows a sample of layout generated with a first version of the Parrot Physical Synthesis Tool [18], using a three-metal layer standard CMOS technology. Such tool was developed at UFRGS (Federal University of Rio Grande do Sul, Brazil). The layouts generated by such tool present transistors in vertical orientation, placed in rows, following the linear matrix style [7]. Some polysilicon connections between the gates of a PMOS-NMOS transistor pairs present doglegs to reduce the misalignment between PMOS and NMOS rows. The lower most metal layer is used for intra cell connections. The second metal layer may also be used for intra cell connections and should run preferably in the horizontal direction. The third metal layer is used for global connections and should run in vertical direction. A transistor sizing tool is integrated into Parrot flow. Transistor sizing is performed in a discrete manner and applied to the layout generation flow of Parrot by the transistor folding technique.

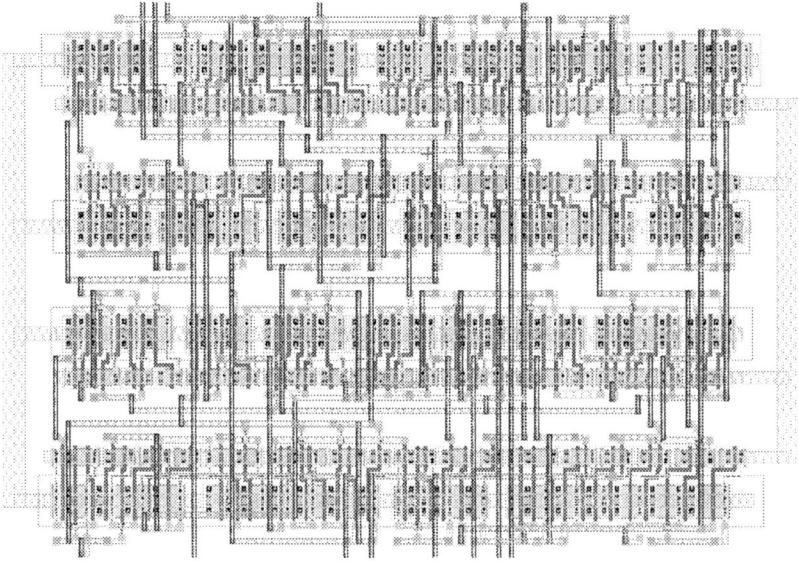

Figure 9. Sample of a layout automatically obtained with the Parrot Physical Synthesis Tool Set [18] developed at UFRGS, where it can be observed that all transistors are vertical and that the gate polysilicon lines have doglegs between PMOS and NMOS transistors. Metal 2 is used to run power lines over the transistors

5.   DESIGN FLOW

Let us start by analyzing what a synthesis methodology is in an abstract fashion. A synthesis system has to accept an input description, which is an optimized and mapped net-list, and output a final layout of the circuit. This is traditionally done by executing a sequence of synthesis steps, carried out by synthesis tools, like placement, global routing, feed-through insertion, detailed routing, transistor sizing and so on. It seems reasonable to break the synthesis down into these steps because they are very different from each other, and each one of them is easier to understand as far as the others are defined. However, they are not independent, and the dependencies they have cannot be linearly decomposed as to form a clean chain that goes from the net-list to the layout without feedbacks and cycles in general. In practice, many systems recognize those loops as single or multiple-pass iterations that are needed to completely solve the problem (or approximate a good solution, as it is NP-Hard). Nowadays, the convergence of the process seems to be even harder to obtain, and most of good recent CAD systems for physical synthesis aim to integrate many different steps and algorithms in a tight loop that refines the solution as best as it can. These tools offer an integrated external view, where for example, placement, global routing, some logic synthesis, re-buffering, are all carried out simultaneously. As they were implemented in the same environment, are tightly integrated, and run over a common database of information (nets, gates and their performance values) that is refined over and over, they are able to converge into an acceptable solution much more than the traditional scripts can. This process often requires incremental algorithms, which can perform partial tasks in partial solutions without destroying or changing previous decisions. Algorithms that behave like this can be invoked both automatically and in the ECO mode (Engineer Change Operation), in which the user wants to make something that she/he sees and will not be reached by the tool alone. In most of the cases, despite all this integration, from the internal perspective they are still sets of algorithms that perform different tasks, running one after the other, what brings us back to the methodology problem. As we see, the methodology problem can be viewed as: "How to break down a complex task into smaller pieces, in such a way that they lead to a global solution when individually solved". Since most of the steps are already well characterized and established for long, our practical and constrained question regarding only the methodology aspect ends up as: "How to sort and link a set of synthesis steps, in such a way that we minimize dependency loops and maximize convergence". The problem must be addressed by someone and re-though from time to time if we want to take the most from CAD tools. Just some basic guidelines help in this context, which we find useful for both the whole synthesis flow and for the routing part of our particular synthesis methodology.

5.1   Generic Methodology Problems

There are two significant generic situations that arise when establishing a methodology. The first is the presence of cycle dependencies, and the second one is the generation of problems whose solution will not be possible to obtain. Cycle dependencies happen in many situations. As an example: the correct placement of a cell in a row of a gate array depends on the routing capacity of the surrounding channels, which in turn depend on the placement of this and the other cells. It is a "chicken and egg" problem, and we can represent it as in Figure 10a as a cycle of dependencies between tasks A and B. Whenever this situation happens it would be useful to analyze which dependency is stronger to sort out the tasks as to break only the weaker dependency. Then, it is possible to implement an estimate of the

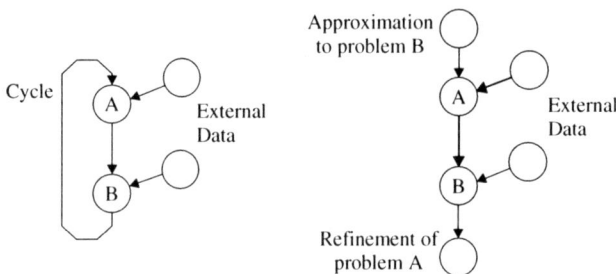

Figure 10. Breaking down cyclic dependencies

solution to B that will be considered by A, and an optimization of A after B is completed. An example of rough estimation is a constructive placement [6].

The other problematic situation relates to the existence or not of a solution for some problem instances. Let be A and B two problems such that each B instance is created by a solution to A, and not all B instances are solvable. Again, this may happen frequently in fixed-die layout approaches such as gate arrays, for example, when the space available for routing, placement or sizing is not enough. Intense research efforts are dedicated to increasing the solution capacity of CAD algorithms, augmenting the solvable set so as to minimize the number of dead ends in the flow. This is represented in Figure 11a, but in some cases there is no such option, because some instances are actually impossible to solve and this is not a limitation of the algorithms. In some cases, we can characterize exactly which instances are solvable and which are not, as in Figure 11b. It is quite rare to be able to implement it. The characterization is usually too time-consuming, because we have to check the feasibility for each instance or partial decision that A is going to take. We may be left only with the two last options. The first is shown in Figure 11c and presents the same guarantee of success of exact characterization, but at the expense of a very limited set of allowed instances. As a consequence, we may be too inflexible in solving problem A, wasting power, area, or compromising the whole solution at all. The last alternative represents a heuristic characterization of the solvable set and is expected to provide equilibrium between the limitation and the number of unsolvable instances generated (Figure 11d). As in many other cases, managing the trade-off between risk and premium is essential to tune a methodology that works.

Let us say one last word about optimality. Many algorithms pay a lot in terms of CPU and even complexity of implementation just to achieve the glorious optimal solution. But each algorithm is just a small piece of the synthesis process. And if there is something that is guaranteed not to happen is to synthesize an optimal circuit. Even if all the algorithms were optimal, we would still not be able to find the optimal circuit, as we broke down the problems and the interactions and interdependencies among them prevent us from finding solutions for each one that lead to the global optimum when each problem is isolated. The point here

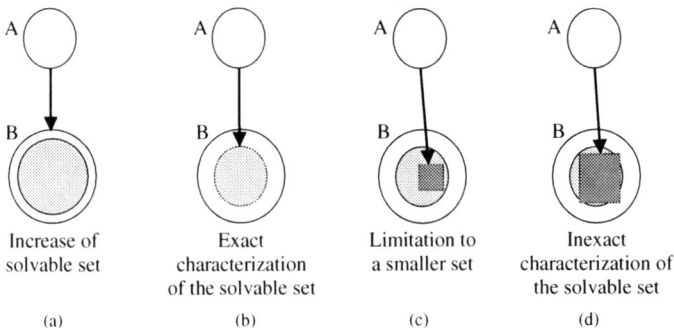

Increase of solvable set	Exact characterization of the solvable set	Limitation to a smaller set	Inexact characterization of the solvable set
(a)	(b)	(c)	(d)

Figure 11. Options for the generation of problem instances that are not solvable

is to identify that the algorithms must be selected and work together with the methodology in order to output a global useful result. Suboptimal algorithms and decisions can be taken once they provide good convergence and guarantee to meet circuit constraints.

5.2   Alternative Methodologies for Layout Synthesis

In traditional physical design flows for random logic generation, given that cells are stored in libraries, the cell design step is conceptually carried out before every other design step in synthesis. One of the possibilities that come out from dynamic cell generation is to move cell synthesis down in the synthesis flow so that it runs after some other synthesis steps. It brings us many potential advantages, like allowing cell synthesis to be aware of its environment, for instance, fan-in, fan-out, routing topology, surrounding layouts and so on, and to simultaneously address area, timing and power optimizations. If cell synthesis was the last step, done after routing, each cell could be generated to drive its actual load, including that caused by the routing. And even if it was not possible to solve severe timing problems that are caused by a bad placement and routing, later cell synthesis could save power and improve yield for all those cell/nets whose time closure was already met. On the other hand, there is a cycle problem if the cells are to be synthesized after detailed routing. It is not simple to make routing with pins whose locations are not previously known, and after that to synthesize and optimize cell layouts that fit into the same fixed layout positions. Therefore let us leave this ideal and difficult flow aside and concentrate on an alternative order of synthesis steps that is neither too traditional nor too revolutionary.

In this alternative flow, depicted in Figure 12, the layout synthesis is split into two steps. The first step is done right after placement and performs transistor sizing (including folding) and ordering for each cell row as a whole. Transistor ordering is based on the same Euler path algorithms for long used in cell synthesis. But instead of analyzing each cell separately, the whole row is fed into the algorithm

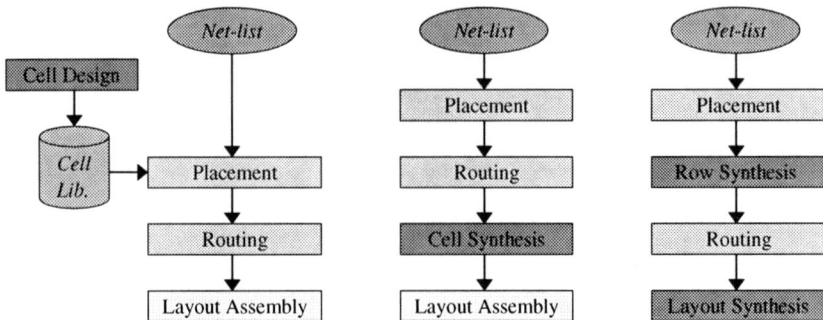

Figure 12. Traditional Standard Cell and new synthesis flows with automatic cell generation after the placement or routing

that tries to minimize diffusion breaks as much as possible, increasing the share of source/drain areas between transistors, which are not necessarily from the same static CMOS cell. In actual instances the number of diffusion breaks turn out to be very low, as if there was only one or a few big cells in each row. In fact, this process does implement cell synthesis aware of its surroundings, although in a limited fashion when compared to the possibilities mentioned above. It has some point of inflexibility in the transistor placement and sizing that does not allow full optimization (or better space usage) in the layout after detailed routing is done. But no matter how limited it is, it already improves over traditional methods in that the layout is more self-adapted, compacted, and this leads to wire-length reduction in general, which has strong correlation to other performance figures as delay and power. Following this row synthesis step, in which transistors and pins are logically placed, runs the routing. Finally, the actual layout is generated accounting for additional spaces, both vertical and horizontal tracks that the routing needed to insert. This last synthesis step resembles an expansion from symbolic to detailed layout, but only at this point we are able to determine exact positions for the transistors, pin access points, and so on, and even the leaf cell's internal routing is affected by the previous steps. That is the reason for calling it layout synthesis instead of just layout assembly as in other methodologies, where cells are just instantiated at the right places and the routing placed over them.

6.   SYNTHESIS TOOLS

This section covers some automation tools that make the core of the layout automation. Placement and Routing tools may employ the same strategies and algorithms no matter they are targeted to standard cell libraries or to custom circuit synthesis, but at least some differences apply. Tools for cell synthesis, on the other hand are very dedicated to the problem, and should implement the layout with the disposition and strategies previously described.

6.1   Placement

Placement algorithms have presented an interesting development in recent years. Both academic and commercial tools went from time-consuming and limited implementations of old algorithms to newer versions that can handle millions of cells in a flat description, being scalable and presenting better performance in terms of area, timing, power and so on. The most significant contributions reside in the development of analytical placers and the assessments of placement solution quality provided by Chang, Cong and Xie. Despite, many other algorithms have undergone significant improvements, some of them enlightened by improvements in their competitors.

Analytical placers work by solving a linear system of x and y coordinates that represents the optimization of a quadratic system of forces. The presence of I/O pins is necessary as input to pull the cells toward the boundaries of the circuit, although most of them still get overlapped close to the center. Spreading the cells was a major concern to the use of this strategy since its first introduction. FastPlace [12] brought the idea back to life by using a smooth spreading strategy that gradually eliminates overlapping while at the same time changing the cost function from the quadratic domain to a linear wire length model that better matches the usual metric of a good placement. Chang et al. have presented Peko [11], a generator of synthetic placement benchmarks whose optimal solution is known, showing that the best placers had solutions 40% worse than the optimal. Although researchers found that Peko benchmarks are unrealistic, having no global connections like real circuits, and that the claimed difference is not that dramatic in real designs, these benchmarks revealed at least some scalability problems of the tools and the importance of the local, short connections in the total WL measure.

Other important aspects that come into account in placement are the amount of free space present in the area, the presence of fixed or movable macro-blocks, and also if the area is fixed or expandable. In [13], Patrick Madden has shown that a Recursive Bisection based placer performs well in general regarding WL if it is given the smallest circuit area, but performs poorly if a larger area with additional spaces is given, because it is based on spreading the cells evenly. On the other hand, FastPlace, as many other algorithms, has to be adapted to work with macro-blocks because they are harder to deal with regarding the overlapping elimination functions [14]. These and other subtle observations are important as real designs use some IPs implemented as macro-blocks and in some cases have as much as 40% of free area to guarantee that the routing will be possible.

Yet another interesting approach for placement and routing was presented in [15]. It was observed that the time-driven placement approaches try to meet timing by optimizing it at the expense of wire length and routability, but if the timing model is not sufficiently accurate, this effort may be ineffective and degrade the solution. In a $0.18\,\mu$m technology the interconnect capacitance per unit can vary up to 35 times. If timing is not met, P&R must be rerun in a loop that is not likely to converge. The approach proposed in [15] do not use timing-driven placement, but let the placement concentrates in a single objective that will be much better optimized alone: reducing wire length. The approach employs variable-die placement and routing, in which spaces are inserted and both placement and routing are adapted as in an ECO mode to solve conflicts, and also uses dedicated cell synthesis where most of the timing optimization can be performed. A significant advantage of the variable-die approach is that every net is routed close to its minimum wire length, which was already well optimized by the placement, and only commercial tools were used to implement a flow that demonstrates it.

6.2   Routing

Actual routing tools need to provide reasonable accuracy. In other terms, if the placement is routable, it has available spaces where needed, the routing tool is expected to make all the connections in such a way that they match previous estimations. It seems not so complicated to achieve such simple goal when the previous steps are already finished. On the other hand, the routing step is the one involved with the biggest amount of information and details. Any local geometrical constraint not handled can potentially let the routing incomplete, making it unable to achieve even its simplest goal of presenting a valid solution. Therefore, we will analyze some approaches considering these two main aspects: completion and accuracy. The routing does also have its own degree of freedom, its capacity of optimizing circuit performance, but we will consider this as a second objective, as the core of the interconnect performance must be provided by the methodology and the placement quality.

Let's consider the four approaches to make area routing represented in Figure 13. The first is a divide-and-conquer technique in which the area is arbitrarily divided into Global Routing Cells (GRCs) and each GRC is routed by a local and restricted algorithm (as a switchbox router or so). This situation leads to the definition of the Cross-Point Assignment problem (CPA) [16] that has to determine the positions at the interface of the GRCs in which the global nets must pass through. One of the main difficulties with this methodology is that there is too little flexibility after the determination of the cross-points, and therefore a solution may be hard to achieve without sacrificing some space and quality. In fact, some approaches to the CPA problem consider entire rows and columns to minimize net crossings, almost as if there were no GRCs at all. And it is almost impossible to insert spaces in the circuit to solve local problems when using this methodology. Yet its potential resides in the isolation of the problems (thus preventing local problems from affecting other GRCs) and in doing the routing in parallel.

The second approach is the most traditional "one net at a time" routing, also called "single net routing". It is characterized by the use of the ubiquitous maze router, which considers the routing of a single net over the whole area, although restricted to pass only through the GRCs that the global routing set as allowed for that net, if a global routing is used. There is no doubt maze routing is a powerful tool to realize the routing of any net. Its main advantages are that it always finds a solution if one exists, and that the path found is optimal. An algorithm with these properties is called admissible. Yet its main disadvantages are huge CPU

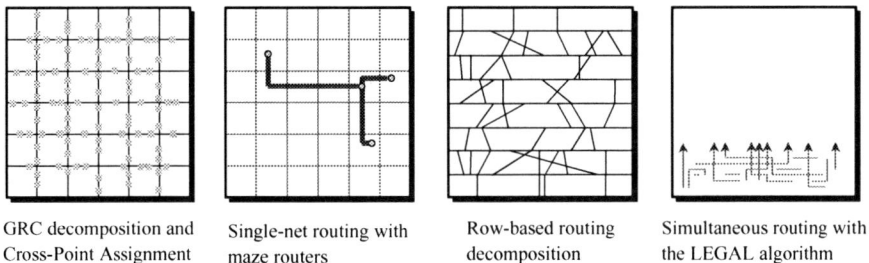

| GRC decomposition and Cross-Point Assignment | Single-net routing with maze routers | Row-based routing decomposition | Simultaneous routing with the LEGAL algorithm |

Figure 13. Four different approaches for Area Routing

running times and the ordering in which nets are routed. The CPU time can be minimized by the correct implementation of an A* in place of a plain BFS search, without sacrificing quality. Yet the ordering problem is much more significant as it can prevent the tool from finding a valid solution, or drastically degrade its quality. When this admissible algorithm is used for a set of nets, there is clearly a greedy approach, in which we chose the optimal solution for a partial problem that does not necessarily lead to a global optimal or to a feasible solution at all. This problem is compensated by the equally known "rip-up and re-route" technique, which implements a loop that gradually solve conflicts and is expected to converge. Even the claimed optimal routing for a single net is relative, due to different figures of merit in performance and path optimization techniques that run after the maze searches. Both speed and optimality aspects of the "single net" approach will be considered in greater detail below.

The third approach represented in Figure 13 derives from the standard cell methodologies, in which the routing is essentially row-based. Although in the standard cell this is a very natural and straightforward idea, it is not a usual approach for Area Routing, due to the area constraint characteristic of the later. Area Routing can be considered as a Fixed-Die approach, in which the circuit must be routed without inserting additional spaces, which is key to the success of standard cell methodology. In this approach the routing is analytically decomposed into a set of fixed-height channels, controlling at the same time channel congestion and network topologies. It implements the same divide-and-conquer idea of the first approach, and minimizes net ordering dependencies, but using less Cross-Points and different algorithms.

The fourth approach that we show here is the most innovative to this problem. It consists on a specific algorithm called LEGAL that is able to simultaneously handle all the connections in Area Routing with much less intermediate definitions. The algorithm is a combination of the Left-Edge and Greedy channel routing algorithms (thus the name) and processes the routing in a single pass over the circuit, as in a scanning process. It does require a previous global routing, and also the insertion of some intermediate net positions that correspond to Steiner Points, to guide its decisions, but is considerably more flexible than the others. The algorithm and its results are also briefly outlined below.

As in many NP-Hard and NP-Complete problems, there might be several other different approaches and algorithms for approximating optimal solutions. For routing we have hierarchical and multi-level algorithms [10], we may divide the problem down into planar routing layers, and so on. Paying attention to the approaches described above does not mean we want to exclude any of these possibilities or any other formulations, but in the rest of this chapter we want to concentrate on the contributions for the last four approaches only.

6.2.1   Path search methods

Maze routers used in the VLSI domain are nothing else than restricted versions of the Breadth-First Search or Dijkstra Shortest-Path algorithms that run on general graphs. Since the beginning, EDA researchers were concerned with implementation aspects. The need for detailed routing motivated several works that reduced the amount of information the algorithm must keep for each grid node to as low as 2 bits. Today it is more than adequate if our grid nodes are implemented with something below 64 bytes for a global routing tool, or a bit smaller for detailed routing, and the use of $\mathbf{A}^*$ (or $\mathbf{LCS}^*$, [17]) instead of plain BFS is straightforward.

$\mathbf{A}^*$ has two main advantages for routing. The first one is the normal economy of nodes expanded by the algorithm. Geometric distance is used as heuristic, what prevents it from expanding too many nodes far from the target. The second advantage is much more pronounced and happens when the costs in the grid are set to unit (or anything uniform over the grid). In this special case, all nodes have the same $f = g + h$ value, for g (cost from source to current node) increases as h (estimate from current node to target) decreases, in the same amount, e.g., the estimates are always perfect. The net result is that only the nodes that reside in the optimal path (or one of them, as usual in a grid) are expanded. All one need to do to get this behaviour is to sort the open queue by f and then by h. Its search pattern is almost like in those old linear-search methods, but with all search capabilities and properties of a shortest path search. Hence, it is important to take advantage of this fact mainly when implementing a detailed routing. Keep the costs uniform as much as possible. Even for global routing, first finding net topologies that are independent from each other may let us use path search methods with uniform costs in surprisingly short run times.

On the other hand, when implementing a 2.5D grid for routing, it is tempting to increase the cost to cross between adjacent mask layers (the third dimension) in order to prevent small connections from using the topmost layers. When doing this, the cost range is abnormally increased (compared to the first two dimensions), and the performance of $\mathbf{A}^*$ will certainly degrade. It will first expand huge areas in the bottom layers before getting to a single node of the layer above them, and this is not what we normally expect from a connection that can be implemented as a straight line, as an example. The big message here is to be aware of the compromise among node cost, path quality and search effort. In this last case of 2.5D grids, a few tricks would help, as taking the expected path length and setting layer dependent constant components of the cost function so as to keep the cost range small and force the nets to prefer some layers over the others according to their lengths.

Despite the great advantages of $\mathbf{A}^*$, all path-search routers suffer from some problems, and we must go through a few of them. There is intrinsic sub-optimality present whenever a weighted graph represents a real world problem. Additionally, in path-search methods, we limit routing layers to some preferred directions, and after that a post-processing is performed to substitute some turns by single layer doglegs [19]. There is no guarantee of optimality under this model (layer constraints and post-processing). It is well known that single net routing also suffers from the net ordering problem, in which a net may become blocked by paths found for nets routed before it. An important observation is that the probability of blockage is inversely proportional to the distance from the net pins, because the number of alternative paths from a node to some place increases as we move farther from this node. Therefore, an interesting solution to the ordering problem is to provide some kind of reserve for nodes close to each circuit pin. This can be implemented as connections to metal layers above cell pins (as pins usually run in lower layers) or in a more general way by means of very limited BFS searches that run from each net pin and stop on all reachable positions two or three nodes away from them. A dynamic reserve can be implemented so that each time one of these reachable nodes must be used by another net, the alternative reachable nodes are checked to see if any is still available so that the net does not get completely blocked.

On the other hand, the availability of a myriad of alternative paths farther from the net pins also brings some difficulty for maze routers. As the circuit becomes congested the router starts to find some very complicated, strange, scenic and extravagant (in terms of resources and power) paths. These paths usually degrade circuit performance and cause much more congestion in regions that were not presenting any congestion problems before. Maze routers also usually ignore congested regions. The graph can embody costs that represent congestion, but tuning the algorithm may be a little difficult. Areas with high cost values will impact all the nets in a similar way, no matter the net is local or global.

When the grid has uniform costs the algorithm will try to scan paths in directions whose priority is determined by the ordering in which neighbours of a node are inserted in the open list. An implementation that is not randomized at this point (and this is not usual) will tend to produce nets that prefer one corner of the circuit over the others. On the other hand, randomizing it only solves the problem with uniform costs, and any small difference in costs (as a single turn in a grid with uniform costs) can make the connection to change from one side to the other of the circuit. The maze router space usage is therefore chaotic and hardly controllable. All of these problems lead to the following overall behaviour: given a circuit that had uncompleted routing, if you estimate the area needed to complete it and insert sufficient spaces in what you think were the right places, after running the maze router a second time it is likely not able to complete the routing again.

6.2.2   Row based routing

One of the possibilities that come out from automatic cell generation is the flexibility on the placement of cell terminals. Once we have this option, it is possible to align cell terminals in such a way that the Area Routing resembles again the old standard cell disposition, e.g., the circuit is logically divided into rows and there are routing channels that appear between the aligned terminals, as in Figure 14. Aligned terminals have been used in a different model – MTM (Middle Terminal Model) – of [8] in a planar fashion for OTC Routing [20] also addressed the use of channel routers in channel-less sea-of-gates for fast prototyping, but comprising a model and algorithms with more constraints than the ATRM and algorithms used herein. The same model of aligned terminals was called CTM (Central Terminal Model) in [21] where a grid-less maze-based channel router try to route channelled circuits making as small use as possible of the channels.

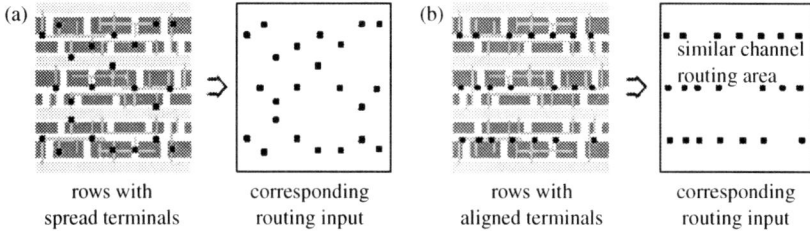

Figure 14. Alignment of pins in Area Routing leads to a channel-like routing

A dedicated routing flow that follows this methodology is implemented in a new router called ChAOS. ChAOS performs the following tasks:
- Channel height balance,
- Virtual pins insertion (feed-through),
- Elimination of overlapped pins,
- Spanning trees computation,
- Channel assignment for net connections,
- Detailed channel routing.

The advantage if this approach resides in its convergence. The algorithms that solve each routing step threat all the connections at the same time, evaluating their needs and inserting small vertical spaces between cells (feed-through) and horizontal spaces between cell rows when they are needed. It does implement a variable-die approach in a similar way old standard cell tools used to do, but considering the particularities of the layout disposition with transparent cells that are implemented in the underlying metal layers.

6.2.3   The LEGAL algorithm

The fact that maze routers present an almost random space usage, ignoring that there are other connections left to route when they process a given one, lead us to think on the way the restricted algorithms like channel routers work. We observed that Greedy and Left-Edge are complementary routing techniques, and that it is possible to combine them and build a new efficient area routing algorithm, called LEGAL [22]. LEGAL processes a symbolic view of the circuit line by line, having as its main control a sequence of steps as in the Greedy algorithm. At each line and step, there are segments that are needed or at least useful for solving the problem. Those that could be implemented are selected in a Left-Edge manner. Figure 15 shows a skeleton of the LEGAL algorithm.

While the horizontal scanning considers the segments in a simple Left-Edge way, the priority in which they must be considered is embedded in the Greedy-like steps. The meaning of this processing is as follows. While each line of the circuit is being routed, previous dependencies are solved, since they are required, and the rest of the space is used to do what can be done in advance. This is a simplified view of the existing priorities. The algorithm is greedy in nature, but is thought to take only local decisions according to clear priorities, and the fact that it considers all nets at

---

Legal algorithm:

```
for each net pin put pin in Terminal[pin.y];
for (line=0; line<circuit.y; ++line)
  {
  extend live columns;
  step1(line); // add new pins
  step2(line); // join split nets
  step3(line); // approximate split nets
  step4(line); // move nets closer to target
  }
```

---

legal_step1:

```
for each pin in Terminal[line]
  {
  calculate target_x; // Left or Right
  found = scan_step1(pin.x,target_x,&found);
  if (not found)
    error("Cannot includ new pin");
  if (Column[found] has no net)
      legal_move(pin.x,found);
  else
    legal_union(pin.x,found);
  }
```

---

legal_step2:

```
for x = 0 to circuit.x
  if Column[x] has a net
    if the net at Column[x] is split
      {
      found = scan_step2(pin.x,target_x,&found);
      if (found)
        legal_union(pin.x,found);
      }
```

Figure 15. Skeleton of the LEGAL algorithm

the same time let it potentially better converge into a solution with 100% routing completion.

Each step performs a specific task: 1- add new pins; 2- join nets present at more than one column (split); 3- approximate split nets; 4- move nets closer to their next terminal. There are separate functions to identify each of these conditions as well as to make the horizontal scanning to find positions that can be used for that specific connection. Figure 14 shows only the step functions for step 1 and 2. Finally, *legal_move* and *legal_union* are utility functions that process horizontal movements and unions of split nets, updating column information and generating the corresponding routing.

In four different implementations of the LEGAL algorithm it has demonstrated to work as expected, although with its particular features and limitations according to each environment. This experiments varied from detailed routing under mask constraints, to unconstrained routing of 2-pin nets, to channel-based global routing. Later works included the definition on how the LEGAL algorithm follows a previous global routing. The global routing, as usual, is needed to define net topologies and distribute congestion, but LEGAL, in particular, must keep all the flexibility it had in deciding the connection paths. This was a challenge that was solved by using a routing entity called half (from half a connection). The half is both independent in respect to its target and transparent in the sense that it can run in the same positions already occupied by the same net. More details of that can be found in [23], and the resulting mechanism clearly separates global and detailed routing without dividing the space into separate parts in this approach.

7. CONCLUSIONS

The paper gives an overview and basic motivation to use physical design automation with synthesis of cells on the fly. Automatic cell generation allows the physical implementation of any logic function to be realized by using Static CMOS Complex Gates (SCCG). Logic minimization using SCCG can reduce the number of transistors by as much as 30%, reducing total circuit area. Area reduction also reduces average wire length, optimizing circuit performance in terms of speed and power. Another important point is that with automatic layout generation the cells could be designed later in the design flow, considering neighboring cells and routing information. In this situation transistor sizing contributes to further improvements in power and time when cells are downsized to reduce input capacitances or enlarged to improve delay. Placement and routing algorithms must reduce average wire length in the first place. But they must also guarantee the routability by the management of routing congestion. The paper has shown some variable-die approaches and methods that contribute on that. For both standard cell and automatic layout generation, the development of tools for pre-characterization is needed. In the past, the performance of the circuit relied on the characterization of cell libraries. As this is no longer sufficient to guarantee circuit performance, new methodologies that rely on automatic cell synthesis can be used and will probably better contribute to get circuits that meet design goals.

REFERENCES

1. SIA International Semiconductor Roadmap, Semiconductor Industry Association (2001).
2. Zenasis, "Design Optimization with Automated Flex-Cell Creation," in Closing the Gap Between ASIC & Custom Tools and Techniques for High-Performance ASIC Design, D. Chinnery and K. Keutzer eds, Kluwer Academic Publishers, ISBN 1-4020-7113-2, 2002.
3. Detjens, E., Gannot, G., Rudell, R., Sangiovanni-Vinccentelli, A.L. and Wang, A. Technology Mapping in MIS. Proceedings of ICCAD (1987) 116-119.
4. Reis, A.; Robert, M.; Auvergne, D. and Reis, R. (1995) Associating CMOS Transistors with BDD Arcs for Technology Mapping. Electronic Letters, Vol. 31, No 14 (July 1995).
5. Reis, A., Reis, R., Robert, M., Auvergne, D. Library Free Technology Mapping. In: VLSI: Integrated Systems on Silicon, Chapman-Hall (1997) pg. 303-314.

6. Hentschke, R.; Reis, R., Pic-Plac: A Novel Constructive Algorithm for Placement, IEEE International Symposium on Circuits and Systems (2003).

7. Uehara, T.; Cleemput, W. Optimal Layout of CMOS Functional Arrays. IEEE Transactions on Computes, v.C-30, n.5, p.305-312, May 1981.

8. Sherwani. Algorithms for Physical Design Automation, Kluwer Kluwer Academic Publishers, Boston (1993).

9. Sarrafzadeh, M., Wang, M., Yang, X., Modern Placement Techniques. Kluwer Academic Publishers, Boston (2003).

10. Cong, J., Shinnerl, R., Multilevel Optimization in VLSICAD. Kluwer Academic Publishers, Boston (2003).

11. Chang, C., Cong, J., Xie M.. Optimality and Scalability Study of Existing Placement Algorithms. ASP-DAC, Los Alamitos: IEEE Society Computer Press, 2003.

12. FastPlace: efficient analytical placement using cell shifting, iterative local refinement and a hybrid net model. International Symposium on Physical Design ISPD 2004. Proceedings. New York: ACM Press.

13. Ameya R. Agnihotri, Satoshi Ono, Patrick H. Madden. Presentation of Feng Shui. International Symposium on Physical Design ISPD 2005. Proceedings. New York: ACM Press.

14. Chris Chu, Natarajan Viswanathan, Min Pan. Presentation of FastPlace. International Symposium on Physical Design ISPD 2005. Proceedings. New York: ACM Press.

15. Vujkovic, M.; Wadkins, D.; Swartz, B.; Sechen, C. Efficient timing closure without timing driven placement and routing. In: Design Automation Conference, Proceedings, 2004. p. 268-273.

16. Kao, Wen-Chung, Parng, Tai-Ming. Cross Point Assignment With Global Rerouting for General Architecture Designs. Transactions on CAD of ICs and Systems. v.14 n.3, 1995. p. 337.

17. Johann, M., Reis, R., Net by Net Routing with a New Path Search Algorithm. 13th Symposium on Integrated Circuits and Systems Design, Manaus, Proceedings, IEEE Computer Society Press (2000) pg. 144-149.

18. Lazzari, Cristiano. Automatic Layout Generation of Static CMOS Circuits Targeting Delay and Power Reduction. Master Dissertation. UFRGS/PPGC. Porto Alegre. 2003.

19. Johann, M.; Reis, R. A Full Over-the-Cell Routing Model. In: IFIP International Conference on Very Large Scale Integration, VLSI, 8, 1995, Chiba, Japan. Proceedings, 1995.

20. Terai, M.; Takahashi, K.; Shirota, H.; Sato, K.. A new efficient routing method for channel-less sea-of-gates arrays. Custom Integrated Circuits Conference, 1994.

21. H. Tseng; Sechen, C.; A gridless multilayer router for standard cell circuits using CTM cells. IEEE Trans. On CAD, Volume 18, Issue 10, Oct. 1999 Page(s):1462-1479.

22. Johann, M.; Reis, R. LEGAL: An Algorithm for Simultaneous Net Routing. In: Brazilian Symposium on Integrated Circuit Design, 14, Pirinópolis, 2001 Proceedings. Los Alamitos: IEEE, 2001.

23. Johann, M.; Reis, R. A LEGAL Algorithm Following Global Routing. In: Brazilian Symposium on Integrated Circuit Design, 15, Porto Alegre, 2002 Proceedings. Los Alamitos: IEEE, 2002.

CHAPTER 6

BEHAVIORAL SYNTHESIS: AN OVERVIEW

REINALDO A. BERGAMASCHI

IBM Thomas J. Watson Research Center, P.O. Box 218, Yorktown Heights, NY 10598, USA.,
Tel: +1 914 945 3903, Fax: +1 914 945 4469, e-mail: berga@us.ibm.com

Abstract: Behavioral synthesis bridges the gap between abstract specifications and concrete hardware implementations. The behavioral synthesis problem consists of mapping a largely
incompletely defined behavior into a well-defined hardware implementation with fixed
sequential behavior. This paper presents an overview of the main algorithms and issues
related to behavioral synthesis and analyzes the pros and cons of each technique. It also
discusses the place for behavioral synthesis in the overall digital design methodology and
the main advantages and drawbacks of its use

Keywords: Behavior, High-level, Synthesis, Scheduling, Allocation, Resource Sharing

1.   INTRODUCTION

Transistor densities in digital circuits (IC's) have increased at approximately
58%/year (compound growth rate), however, designer productivity has grown at
about 21%/year (compound growth, measured in transistor per person-month) (SRC,
1994). These figures have prompted many warnings about an impending design productivity crisis. This is to be caused by the inability of designers to use effectively
all the available transistors in reasonable time and cost. Given that it is infeasible to
continue increasing the size of design teams (from a cost and manageability point
of view), designers have relied more and more on computer-aided design (CAD)
tools to bridge this *productivity gap* between what can be fabricated on a piece of
silicon and what can effectively be designed; the former greatly outweighing the
latter.

One of the most promising ways for bridging this gap is increasing the level of
abstraction at which designs are specified. High-level, abstract descriptions can be
much shorter than register-transfer-level (RTL) and gate-level descriptions for the

R. Reis et al. (eds.), Design of Systems on a Chip, 109–131.
© 2006 *Springer.*

same functionally, size and performance. Therefore, it is much more productive in cost and design time to describe a chip in 50 thousand lines of behavioral or algorithmic code (e.g., VHDL, Verilog) than in 300 thousand lines of gate-level-code; provided that similar performance levels are achieved. Figures from our designs and other industry sources indicate that behavioral descriptions can be from 5 to 10 times shorter than their RTL/gate-level equivalents.

At the same time, users, tool providers and researchers alike cannot quite agree on what exactly a behavioral description is. For a designer who has always used schematic entry at the gate-level, a hardware description language (HDL) at the RT level is very abstract. Similarly, for behavioral synthesis, the input is an algorithmic description and the output is at the RT level. Going further up, the instruction-set model of a microprocessor can be much more abstract than current behavioral synthesis systems accept.

Behavioral synthesis borrows techniques from many disciplines, ranging from languages and compilers, to graph theory, to hardware design. However, these techniques have to be specifically modified in order to be efficient for synthesis. This is because the cost functions or optimization criteria useful for other disciplines may not apply to behavioral synthesis.

For example, some of the early scheduling techniques were borrowed from microcode optimization and software compilation. In these domains, chaining of operations is not applicable since the units being optimized are microcode words or instructions. In behavioral synthesis this is not the case since chaining of gates and functional units may be desirable.

Another limitation of early systems was the limited or no support for conditional operations. This, again, was a legacy from early software compilation techniques which could only analyze program graphs on a basic-block basis. This limitation was not a problem for early behavioral synthesis systems because they were targeted towards data-processing applications, such as digital-signal processing (DSP), which contained no conditional operations. As the application domains were expanded to control-dominated designs, these techniques proved inefficient and new control-flow-based algorithms were developed.

When compared to other hardware design techniques, such as logic synthesis, behavioral synthesis has one main difference. That is, it allows for changes to the sequential behavior of a specification. The task of scheduling may change the cycle-by-cycle behavior of the specification, and this has overwhelming repercussions on the overall design and verification methodology.

Behavioral synthesis bridges the gap between abstract language specifications and concrete hardware implementations. The behavioral synthesis problem consists of mapping a largely incompletely defined behavior into a well-defined hardware implementation with fixed sequential behavior.

This paper presents an overview of the main algorithms and issues related to behavioral synthesis and analyzes the pros and cons of each technique. It also discusses the place for behavioral synthesis in the overall digital methodology and the main advantages and drawbacks of its use.

2.   ABSTRACTION LEVELS

There are three abstraction levels commonly associated with synthesis tools: behavioral, register-transfer and gate-level. These levels, however, are not totally disjointed, and it is very common to mix in the same specification parts described at the behavioral level with RTL constructs. This mixed use of different levels in one design has led to common misconceptions when describing the capacities of synthesis systems. Figure 1 illustrates these abstraction levels and their characteristics.

Region 1 covers gate-level netlist descriptions containing Boolean equations, gates and registers. Synthesis of these netlists requires gate-level technology independent optimizations and technology mapping.

Region 2 is termed RTL netlist description because it adds RTL modules to gate-level netlist of Region 1. These RTL modules are multi-bit primitives such as, adders, multiplexes, comparators, etc. Synthesis of this level requires, first, that these RTL modules be expanded into simpler logic gates, or mapped to existing library elements directly. For example, a 32-bit adder may be expended into gates using a carry-save-adder algorithm and then optimized jointly with the rest of the logic. Alternatively, it may be directly mapped to an existing adder box picked

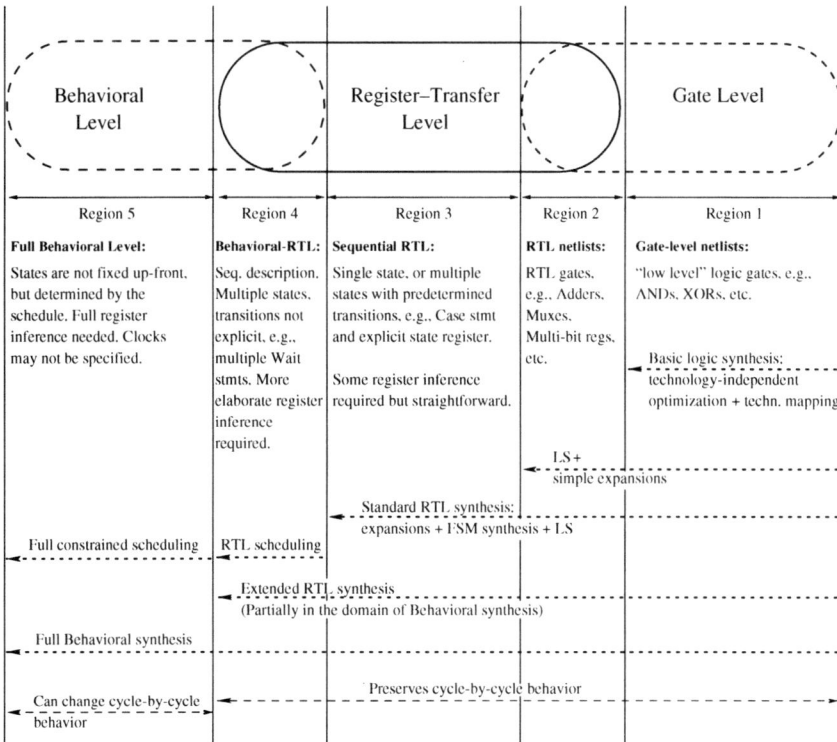

Region 5	Region 4	Region 3	Region 2	Region 1
Full Behavioral Level:	**Behavioral-RTL:**	**Sequential RTL:**	**RTL netlists:**	**Gate-level netlists:**
States are not fixed up-front, but determined by the schedule. Full register inference needed. Clocks may not be specified.	Seq. description. Multiple states, transitions not explicit, e.g., multiple Wait stmts. More elaborate register inference required.	Single state, or multiple states with predetermined transitions, e.g., Case stmt and explicit state register. Some register inference required but straightforward.	RTL gates, e.g., Adders, Muxes. Multi-bit regs, etc.	"low level" logic gates, e.g., ANDs, XORs, etc.

Figure 1. Abstraction levels of hardware descriptions used in synthesis

up from a library of elements. Region 1 and 2 deal with combinational logic plus registers, with no concept of finite-state-machine (FSM) states.

Region 3 is the most common RTL description currently supported by logic synthesis. It usually involves a sequential description where registers may not be explicitly declared, but must be directly inferred. FSM states, if used, must be explicitly declared, and all state transitions clearly specified. Synthesis at this level has to deal with very simple register inference, as well as FSM optimization and synthesis. Most current logic synthesis tools can handle this level.

Region 4 starts invading the behavioral-level domain. This description is sequential, possibly with multiple states, but the transitions are not explicitly declared and have to be derived. It also requires a more elaborate register interface scheme, where lifetime analysis is required. Synthesis in this region requires some behavioral techniques, such as, limited scheduling for deriving the required FSM, also called RTL-scheduling (Bergamaschi et al., 1995) or Fixed-I/O scheduling (Knapp et al., 1995), as well as data-flow analysis systems for register inference. Until around 1997, few commercial synthesis systems could handle this level.

Region 5 covers full behavioral level, which is a sequential description with or without a specified clock or FSM states. Registers are not pre-specified, nor can they be directly inferred. This requires full behavioral synthesis in order to define the states (through scheduling), map operators to gates and RTL modules and map variables to registers.

Figure 2a shows an example of a behavioral description (region 5) of an algorithm for computing the greatest-common divisor of two numbers. In this example, there is no implied clock and there are no implied states or registers. It is the task of the behavioral synthesis tool to implement it, subject to user constraints. Figure 2b shows the same algorithm written in a slightly lower level (region 4). In this case, the state boundaries are clearly defined by *Wait* statements, but the states and transitions need to be derived by synthesis tool. There are no explicit registers, but given that the states are defined, all registers can be inferred through data-flow analysis. Figure 3 presents the same algorithm as an explicit finite-state machine plus data-path model. The state-transition logic is explicit given in the form of a *Case* statement in Process **PO**, and the next-state assignment, and the data-path registers and operations are described in Process **P1**. Although there are no explicit registers, the inference is direct since **P1** is process with *Wait* statements and all operands are declared as *signals*.

What characterizes a synthesis tool as behavioral or RT level is not necessarily the level of the input description that it can handle but mainly the algorithms that it contains. For example, one could perform scheduling on a purely RTL description and create extra states due to constraints – this would consist of a behavioral synthesis transformation although the input was plain RTL. Moreover, a synthesis tool that operates in a certain region, must be able to handle descriptions below that level. So designers can easily mix different description levels in the same design and use a single tool to process it.

(a)

```
Entity GDC_BEH is
Port (rst : in bit;
        xi, yi : integer range 1 to 65535
        val : out integer range 1 to 65535);
end GCD_BEH;

Architecture Behavior of GCD_BEH is
begin
  P1: Process (xi, yi, rst)
    variable x, y : integer range 1 to 65535;
  begin
    if (rst = '0') then
      x := xi; y := yi;
      while (x /= y) loop
        if (x < y) then y := y x;
        else x := x y;
        end if;
      end loop;
      val <= x;
    else
      val <= 0;
    end if;
  end process;
end Behavior;
```

(b)

```
Entity GDC_BEHRTL is
Port (clk : in bit; rst : in bit;
        xi, yi : integer range 1 to 65535;
        val : out integer range 1 to 65535);
end GCD_BEHRTL;

Architecture Beh_rtl of GCD_BEHRTL is
begin
  P1: Process
    variable x, y : integer range 1 to 65535;
  begin
    wait until not clk'stable and clk = '1';
    if (rst = '0') then
      x := xi; y := yi;
      L1: Loop
        wait until not clk'stable and clk = '1';
        if (x = y) then Exit;
        elsif (x < y) then y := y x;
        else x := x y;
        end if;
      end loop;
      wait until not clk'stable and clk = '1';
      val <= x;
    else
      val <= 0;
    end if;
  end process;
end Beh_rtl;
```

Figure 2. GDC algorithm written in: (a) full behavioral VHDL, (b) hybrid behavioral-RTL VHDL

3.   BEHAVIORAL SYNTHESIS TASKS

Behavioral synthesis includes a number of algorithmic tasks, which convert a language or graphical specification of a design into an implementation as an RT/gate-level netlist. These algorithmic tasks span the domains of programming language compilers, graph theory, operations research, and hardware synthesis. This section contains an overview of the most important techniques and algorithms used.

3.1   Language Parsing and Data Model Generation

This is the very first step in all language-driven behavioral synthesis system. It comprises of parsing the input specification and performing syntactical and semantical checks. The output of this process is the generation of a graph representation of the specification, which is normally called *Control and Data Flow Graph* (CDFG). This step is very important because it must capture all the information present in the specification, which can be very extensive if general purpose HDL's, such as VHDL and Verilog, are used. For example, the data model must be able to represent

```
Entity GDC_BEHRTL is
Port (clk : in bit; rst : in bit;
       xi, yi : integer range 1 to 65535;
       val : out integer range 1 to 65535);
end GDC_BEHRTL;
Architecture Beh_rtl of GCD_BEH is
   type GCD_States is (START, COMPUTE, OUTPUT);
   signal cur_state, nxt_state : GCD_States := START;
   signal x, y : integer range 1 to 65535;
begin
```

```
                                           P1: Process
   P0: Process (xi, yi, rst)                begin
   begin                                       wait until not clk'stable and clk = '1';
      Case cur_state is                        Case cur_state is
         when START =>                            when START =>
            if (rst = '0') then nxt_state <= COMPUTE;    if (rst = '0') then x <= xi; y <= yi;
            else nxt_state <= START; end if;             else val <= 0; end if;
         when COMPUTE =>                           when COMPUTE =>
            if (x /= y) then nxt_state <= COMPUTE;     if (x /= y) then
            else nxt_state <= OUTPUT; end if;             if (x < y) then y <= y x;
         when OUTPUT =>                                  else x <= x y; end if;
            nxt_state <= START;                        end if;
         end case;                                when OUTPUT => val <= x;
      end process P0;                          end case;
                                              cur_state <= nxt_state; state assignment
   end Beh_rtl;                             end process P1;
```

Figure 3. GCD algorithm written in sequential RTL VHDL

arrays, complex types (e.g., records), synchronous and asynchronous behaviors, and basically every language construct that the designer wants to use. Failure to do so has jeopardized many a commercial synthesis system in the past.

The control and data flow graph generated by this task represents the sequencing of control as well as the flow of data implied in the specification. Although several variations of this graph exist, the main difference consists of whether it uses a joint CDFG or separate graphs, one for control and one data. In joint CDFG's, a node represents an operation and there are two different types of edges, to represent the data exchanged between operations, and control sequence. Examples of joint CDFG's can be found in McFarland 1978, and Parker et al., 1986. Separate graphs are used by Bergamaschi et al., 1995, Camposano and Tabet, 1989, and Lis and Gajski, 1988.

Figure 4 shows the control and data flow graphs for the GDC example in Figure 2b, as used by the HIS system (Bergamaschi et al., 1995).

3.2   Compiler Transformations

These transformations are akin to those performed by programming languages compilers, and aim at simplifying the graph and removing redundancies based on static analyses of the control and data dependencies. Aho et al., 1986 contains extensive descriptions of these techniques. The main transformations applicable to

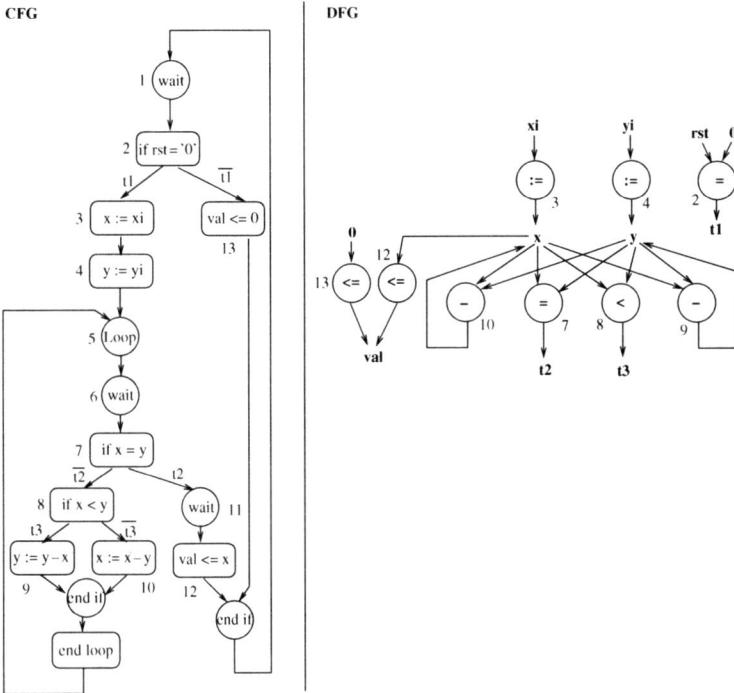

Figure 4. Control and Data-Flow graphs for GCD example in Figure 2b

behavioral synthesis are: dead code elimination, constant propagation, code motion, in-line expansion, loop unrolling, and tree height reduction.

Dead code elimination and constant propagation are always useful since they simplify the graph without any side effect on the synthesis result. Most of the other transformations do affect the synthesis results and should only be performed if appropriate.

Code motion involves moving operations across control boundaries, such as conditional operations and loop boundaries. By moving an operation which is originally in a conditional branch to a point before the conditional operation, makes it independent on the condition being tested, thus possibly reducing the control path to that operation. However, the operation will no longer be *shareable* with other operations in the complementary conditional branches.

In-line expansion brings functions and procedures in-line by flattening them wherever they are called. Normally, sequential procedures can be complete FSM's, which are called from other FSM's in a hierarchical manner. Behavioral synthesis normally operates on one FSM at a time, hence optimizations across procedure boundaries are not performed. One way to avoid this is by in-lining the procedure into the calling FSM and synthesizing it as a flat FSM. While this can lead to more parallelism and optimizations, it also leads larger CDFG'S, which may pose complexity problems for many algorithms, well as being more difficult to debug the final implementation.

Loop unrolling poses similar trade-offs. Loops with constant bounds can always be unrolled. By unrolling a loop, the body of the loop gets replicated, which enlarges the number of operations and allows more optimizations, while increasing run time.

Tree height reduction is applied to chains of operations to reduce the number of operations on the critical path and increase parallelism. While this can be useful for reducing the critical path, it may increase the number of resources required given the increase in parallelism.

In summary, although these transformations deal mostly with language constructs and are applied to the CDFG, they may have significant effects on the amount and types of optimizations that can be performed later on. These effects need to be carefully estimated for efficient use of these transformations.

3.3   Data-Flow Analysis

Data-flow analysis is another technique used in programming language compilers (Aho et al., 1986) which is specially important in behavioral synthesis. The main purpose of data-flow analysis is to compute the *lifetime* of values. Lifetime analysis looks at all operations that assign and use the value of a given variable, also known as the variable *definition-use* chain. The definition-use chain of a value (being assigned to a variable) tells exactly when the value is created and when it is used for the last time, which defines its lifetime.

Behavioral synthesis uses data-flow analysis for three purposes:
1. to *unfold* variables in the data-flow graph (Thomas et al., 1990; Camposano and Rosenstiel, 1989);
2. to determine the values alive at each operation, which is important for deriving the required interconnections and determine which values need to be stored in registers (Bergamaschi et al., 1995) and
3. to derive the lifetimes of registers during resource sharing (Kurdahi and Parker, 1987; Bergamaschi et al. 1995).

Most systems apply data-flow analysis to basic blocks only (Thomas et al., 1990). The HIS system (Bergamaschi et al., 1995) applies it globally, through conditional operations and loops.

3.4   User Constraints

The ability to specify and implement user-defined constraints is very important in behavioral synthesis, from a practical point of view. One of the arguments normally held against behavioral synthesis is that designers find very difficult to control and predict the synthesis result. This is required because designers want to have the option of dictating exactly what the logic should look like in situations where synthesis cannot meet the timing or area requirements. Moreover, if the result of synthesis is unpredictable it becomes very difficult to understand and debug it, even if requirements are met.

Therefore it is important to designers to specify constraints which should be used by the synthesis algorithms. Constraints can be used to specify:

- **Resource constraints**. For example, the number and type of functional units that each design can use. These constraints can be used during scheduling and resource sharing. Maha (Parker et al., 1986), Force-directed scheduling (Paulin and Knight, 1989), and HIS (Bergamaschi et al. 1991b; Raje and Bergamaschi, 1997), are among the many systems that can handle these constraints.
- **Timing constraints**. These constraints have been used in many ways. Nestor and Thomas, 1986 used timing constraints on the interfaces to control how the interface events were scheduled. Kuehlmann and Bergamaschi, 1992 used timing constraints to specify the maximum clock period. Paulin and Knight, 1989 used a timing constraint to specify the maximum number of cycles that the schedule should use.

3.5   Scheduling

Scheduling consists of the assignment of operations (in the CDFG) to controls steps, possibly given constraints on area and/or delay, while minimizing a cost-function. It defines the states and transitions of the controller. Moreover, by fixing the state bounds, it also defines all the values that need to be stored in registers (see Section 3.6).

Figure 5 presents a few important basic concepts applicable to most scheduling algorithms. It shows different configurations of data-flow operations and simple scheduling schemes. The horizontal lines denote scheduling boundaries.

Figure 5a shows a simple sequential schedule, where each operation in a data-dependent chain is placed in a separate control step. By allowing chaining of operations, as in Figure 5b, the schedule can be shortened at the expense of longer cycle times.

If the delay of an operation exceeds the target cycle time, there are two basic options: use a multi-cycled unit (Figure 5c) or a pipelined unit (Figure 5d). A multi-cycled unit is purely combinational, so new inputs can only be taken in every N cycles, where N is the number of cycles that the unit takes to complete. A pipelined unit contains internal storage, so new inputs can be taken in every I cycles, where I is the data–initiation interval of the pipelined unit.

Figure 5e shows a schedule, which can parallelize operations that are not data or control dependent. By parallelizing operations, the schedule can get shorter at the expense of using more resources.

Figure 5f and Figure 5g show the two simplest forms of scheduling: As-Soon-As-Possible (ASAP) and As-Late-As-Possible (ALAP). The ASAP algorithm compacts the operations in states from the top down, respecting the data dependencies. The ALAP algorithm is similar except that the compaction is bottom-up. By not allowing chaining, these algorithms give a very crude measure of the critical paths in number of states as well as the critical states in resource utilization.

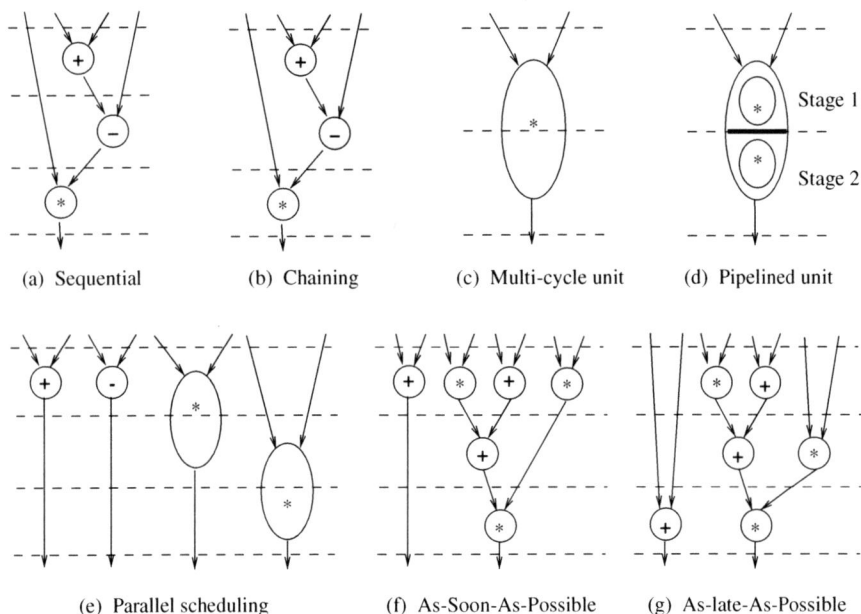

(a) Sequential   (b) Chaining   (c) Multi-cycle unit   (d) Pipelined unit

(e) Parallel scheduling   (f) As-Soon-As-Possible   (g) As-late-As-Possible

Figure 5. Basic scheduling concepts

3.5.1   Data-flow vs. control-flow approaches

Scheduling algorithms have traditionally been developed for two different types of designs: data-flow dominated (e.g., digital signal processing applications) and control-flow dominated (e.g., controllers, FSMs).

Most scheduling algorithms are data-flow-based (DF-based), following the fact that most early applications of behavioral synthesis where DSP algorithms. The most significant DF-based-based algorithms are list scheduling (Davidson et al., 1981; Parker et al., 1986; Pangrle and Gajski, 1986) and force-directed scheduling (Paulin and Knight, 1989; Cloutier and Thomas, 1990; Verhaeg et al., 1995).

DF-based algorithms use the data-flow graph to analyze the data dependencies among operations and schedule groups of parallel operations in consecutive control steps. The goal of these algorithms is either to minimize the total number of control steps under resource constraints or to minimize the resources required under timing constraints (e.g., a maximum number of control steps). DF-based algorithms are very good at exploiting parallelism among operations, however, they are less suitable for handling complex control dependencies and chaining of operations.

As the application domains broadened, scheduling algorithms started to be applied to control-intensive designs, which called for control-flow-based (CF-based) techniques. CF-based algorithms exploit primarily the control dependencies among operations, by analyzing the control-flow graph (CFG) and deriving control conditions for each operation. These algorithms are less suitable for exploiting parallelism because the operation order is dictated by the control flow. On the other hand, they are intrinsically able to determine whether operations are mutually exclusive (by checking their control conditions) and can be scheduling in the same state and share resources. The main goal of CF-based algorithms is to minimize the number of control steps under resource constraints.

The main CF-based algorithm is the path-based scheduling algorithm (Camposano, 1991; Bergamaschi et al. 1991b). Other interesting approaches have been implemented in Amical (Jerraya et al., 1993), PBSS (Wolf et al., 1992), and Olympus (Ku and De Micheli, 1992).

Mixed approaches have also been attempted. Bergamaschi and Allerton, 1988 presented a CF-based algorithm with special ways for parallelizing operations. Wakabayashi and Tanaka, 1992 presented a DF-based algorithm with special treatment of conditional operations.

More recently, Bergamaschi and Raje, 1997a presented an algorithm called ADAPT which combines CF and DF-based techniques for efficient handling of control dependencies and mutual-exclusiveness, as well as data-dependencies and parallelism extraction.

3.5.2   Time-constrained scheduling algorithms

Time-constrained algorithms minimize the number of resources required to schedule a design under a fixed (constrained) number of clock cycles. This is usually useful when the application domain is bound by real-time constraints, such as sampling rates and throughput, as it is the case in DSP applications.

The main approaches in this category are Force-Directed scheduling (FDS) and Integer-Linear Programming (ILP).

Force-directed scheduling (Paulin and Knight, 1989) is a heuristic algorithm, which aims at balancing the resource utilization across all control steps, thus minimizing the overall resources. Given a fixed number of control steps, this algorithm computes the probability of each operation being assigned in each control step. Then it computes a *force* between an operation and a control step, based on the sum of the probabilities of all operations of the same type in the control step. This force indicates the *attraction* or *repulsion* of a given operation to/from a control step. The algorithm computes these forces between each operation and each control step and iteratively selects the operation and control step pair which presents the least (repulsion) force, and schedules the operation onto the control step. It then recomputes the forces and repeats the procedures until all operations are scheduled. FDS is a global heuristic in both selecting the operation and selecting the control step to schedule it. The complexity is $\sigma(cN^2)$, where c the number of control steps and N is the number of operations.

Integer-linear programming approaches are optimal approaches, which minimize a given cost function subject to a number of constraints. The cost function is usually the total cost of the operations, e.g., $\sum_{k=1}^{m}(C_k x N_k)$, where C_k is the cost of operation k and N_k is the number of operations of the same type, for all operation types needed in the design. The constraints are written in the form of equalities or inequalities representing various characteristics, such as data-dependencies and maximum number of functional units of a given type. The ILP solver assigns values to the ILP variables, which minimize the cost function subject to the constraints. The ILP variables indicate in which control step a particular operation is scheduled. The optimality of this approach results in exponential complexity on the number of variables, which can grow very rapidly even for modest size designs, thus severely limiting the practicality of this approach. The main systems using these algorithm are Hafer and Parker, 1983; Hwang et al., 1991 and Gebotys, 1993.

3.5.3   Resource-constrained scheduling algorithms

Resource-constrained algorithms minimize the number of control steps required to schedule a design under a fixed (constrained) number of resources. The constraint is usually expressed as the maximum number of functional units that the design can use, which roughly translates into silicon area bounds. This is a very inaccurate constraint on the silicon area because it does not consider many other area-consuming items, such as the controller and interconnections.

The main example of this type of algorithms is *List Scheduling* (Davidson et al., 1981; Parker et al., 1986; Pangrle and Gajski, 1986) which is one of the most common scheduling approaches. As the name indicates, this algorithm starts by creating a sorted list of all *ready* operations, which are the operations whose inputs are either primary inputs or have already been scheduled. Then it selects the first operation in the list and places it on the current control step, provided that it does not violate a resource constraint. The list may or may not be updated and re-sorted after each assignment. When no more operations can be scheduled on the current control step, a new control step is initiated and the process it repeated until all operations are scheduled.

The sorting of the *ready* list is critical to the quality of the algorithm. The list is sorted according to a priority-function, which attempts to minimize the number of control steps needed. As an example, the priority-function used in Maha (Parker et al., 1986) uses the range of possible control step assignments for each operation as sorting criteria; where range is the interval in control steps between the ASAP and ALAP schedules for the operation. By always scheduling operations with the smallest range (freedom) first, the algorithm prioritizes the operations on the critical paths, which helps reduce the total schedule length. Other priority functions have been proposed by McFarland, 1986 and Pangrle and Gajski, 1986.

List-scheduling is a local heuristic algorithm, which minimizes the number of control steps while meeting resource constraints. The complexity is linear on the number of operations (excluding the sorting step).

The force-directed list-scheduling algorithm (Paulin and Knight, 1989) is a hybrid approach that uses the concept of force (from force-directed scheduling) as the list priority function. Its results are potentially better than those of plain list schedulers due to the more global characteristics of its priority function.

3.5.4   Special scheduling formulations

There are a number of other scheduling algorithms which have a more general formulation for handling constraints, conditional operations and chaining of operations.

The path-based scheduling algorithm (PSB) (Camposano, 1991; Bergamaschi et al., 1991a; Bergamaschi et al., 1991b) was one of the first to handle constraints in a full control-flow-based approach. PSB is a technique, which analyzes all control sequences of operations (paths) in the control-flow graph and schedules each path independently, subject to constraints. These constraints are general restrictions on the schedule and may be due to resources, delay or any other cost measure. The formulation of the PBS algorithm is purely based in paths and constraints, and is not concerned with parallelism or chaining.

The first step of the algorithm is to derive all paths and constraints applicable to each path. Figure 6a shows a simple CFG with one conditional operation and two paths (Figure 6b). Constraints are represented as *intervals* in each path, and indicate that a constraint violation is present. For example, consider the following constraints to be imposed on the schedule:

1. there should be only one adder in the final implementation, and
2. the target cycle time should not exceed 15 ns.

These constraints are mapped as intervals between the operations violating them. Constraint $c1$ in Figure 6b states that operations 2 and 5 cannot be schedule in the same state, since that would require 2 adders, and would violate the constraint. Similarly, when applied to path 2, this constraint creates intervals $c3$ and $c4$. Assuming that operations 4 and 6 are data-dependent and their chained delay

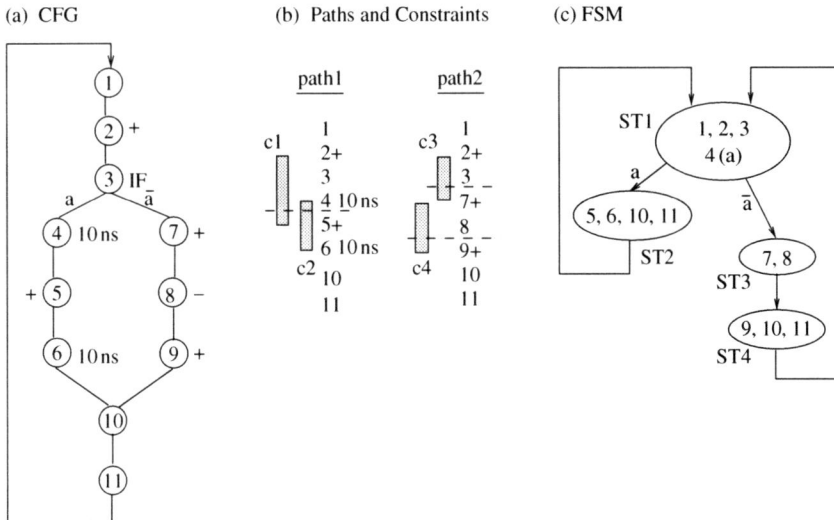

Figure 6. Example of path-based scheduling; (a) CFG. (b) Paths and constraints. (c) Final scheduled FSM

exceeds 15 ns, then constraint interval $c2$ is created to represent constraint 2. In order to satisfy this constraint, operations 4 and 6 must not be chained (scheduled) in the same state.

The PBS algorithm uses clique-covering techniques (Papadimitriou and Steiglitz, 1982) to schedule each path in the optimal (minimal) number of control steps satisfying all constraints. Scheduling steps are represented as horizontal lines Figure 6b. Contrary to DF-based schedulers, not all paths are scheduled in the same number of control steps. DF-based algorithms schedule all paths in the same number of steps, as dictated by critical path; hence shorter paths will still take as long to execute as the critical path. The PBS algorithm schedules each path independently in its minimum number of steps. Figure 6c shows the final scheduled FSM, showing the two paths, one taking 2 cycles (states) and the other taking 3 cycles to execute.

The generality of PBS comes at a price in complexity. The complexity is proportional to the number of paths, which can grow exponentially with the number of conditional operations. This renders it impractical for designs with complex control.

The ADAPT algorithm (Bergamaschi and Raje, 1997a) contains a hybrid formulation of PBS, with parallelism extraction and partitioning, which is capable of avoiding run-time complexity problems by partitioning and trading.

Other special scheduling formulations have been presented by Wolf et al., 1992, and Ku and De Micheli, 1992, among others.

3.5.5   *Domain specific formulations*

Specific scheduling formulations have been proposed to deal with digital-signal processing designs. These designs are characterized by data-flow operations, commonly described in a loop, where values computed in one iteration are passed along to successive iterations. In this context, the goal of scheduling is to increase throughput of the loop, or obtain a given throughput, while minimizing resources. Throughput is dictated by the rate, at which new samples arrive and have to be processed. Loop pipelining is a technique whereby multiple iterations of the loop are overlapped in pipelined manner, thus allowing multiple data samples to be processed by the loop hardware at the same time. Different loop pipelining techniques have been presented by Goosens et al., 1989; Gebotys, 1993 and Passos et al., 1994.

3.6   Allocation and Resource Sharing

Allocation consists of mapping abstract operands, operations and data transfers, present in the data-flow graph (DFG) onto registers, nets, functional units, multiplexer and busses. It defines the data-path of the design. Allocation can be subdivided into three main tasks:
- **Storage Binding**. It maps data values (variables, signals) present in the DFG onto storage elements (e.g., registers, register-files, memories) in the data-path. Storage elements may be created for several reasons. In a general RTL sense, registers may be created for variables and signals according to the language

register-inferencing rules. In a behavioral sense, registers are needed to store values, which are created in one control step and used in another.

- **Functional Unit Binding**. It maps operations in the DFG to functional units in the data-path. All operation types present in the DFG, e.g., logic gates, adders, multipliers, need to be mapped to functional units. This mapping need not be a one-to-one relationship, as multiple operations in the DFG may be mapped to the same data-path unit. Moreover, the binding can be selective; for example, an addition in the critical path may be bound to a carry-look-ahead, whereas a less critical addition may be bound to a ripple-carry adder to save area.

- **Interconnection Binding**. It maps data-transfers in the DFG to interconnection paths in the data-path. Data-transfers take place whenever an operand is used as input of an operation or assigned as output of an operation. Interconnection paths are represented by multiplexers and busses (which are selected by control logic). For example, a simple operation "$a <= b + c$" may require a sequence of interconnection paths to be switched on, depending on the interconnection scheme (busses or multiplexed point-to-point) adopted. First, the value of b may have to be retrieved from a register file, placed on the bus and stored in a register at the first input of the functional unit implementing the addition. Then, in the following cycle, the value of c has to follow the same path and be stored in a register at the second input of the functional unit. With both inputs ready, the addition can take place and the result can be put back on the bus and stored as a onto the register file.

Allocation algorithms usually follow two general approaches: constructive, and global decomposition. In both approaches, the algorithms need to know which resources can be shared. For example, it needs to know if two values (that need to be stored) can be mapped to the same register, or if two additions can be mapped to the same adder. Thus, before proceeding to the description of the algorithms, it is important to learn the conditions under which resources can be shared.

Conditions for resource sharing vary according to the element to be shared, and are different for registers, functional units and interconnections.

- **Register sharing conditions**. Values in the DFG, which need to be stored, can be stored in the same register if and only if the lifetime intervals of the values do not overlap. Based on the scheduled states and on the lifetimes computed by data-flow analysis, one can derive the lifetime intervals for stored values and determine if they overlap.

- **Functional Unit sharing conditions**. Two or more distinct operations in schedules DFG can be implemented by the same functional unit if and only if the operations are never executed simultaneously. This is the case when operations are scheduled in different states or executed under mutually exclusive control conditions.

- **Interconnection sharing conditions**. Multiple data-transfers can be implemented over the same bus if and only if the values are mutually exclusive in time (on a clock cycle basis). In the interconnection binding example previously described, the values of both b and c are brought to the functional unit over the same bus into two cycles; that is, the data-transfers although required in the same operation, can be scheduled in two different states, thus allowing bus sharing.

3.6.1   Constructive allocation approaches

In a constructive approach, the data-path is formed in a step-by-step fashion, where at each step it chooses to perform one of the binding tasks describe above. The biding selection is based on a cost-function which determines the data-flow value, operation or data-transfer that should be mapped in that step and whether it should be mapped to an existing data-path element (register, functional unit or multiplexer), or a new element should be created.

The cost function that selects the value, operation or data-transfer to be bound is usually based on area or timing, for example, selecting the candidate that would add the least amount of area to the data-path being formed. This is a local heuristic, which may produce inefficient results as the number of operations, and binding possibilities grow. Examples of systems that use this approach are Emucs (Hitchcock III and Thomas, 1983) and Mabal (Kucukcakar and Parker, 1990).

3.6.2   Global decomposition allocation approaches

These approaches decompose the allocation and resource sharing problem into independent tasks for each of the binding steps. Each task is formulated as graph-theoretical problem and solved using existing techniques.

Contrary to the constructive approaches where the binding steps for registers, functional units and interconnections were interleaved in a greedy manner, global decomposition approaches solve each task separately; however, since they are interdependent, there is no guarantee of an optimal solution even if each task is solved optimally.

Tseng and Siewiorek, 1996 were the first to present a graph-theoretical formulation to the allocation and resource sharing problem. According to their formulation, the three binding tasks were modeled in the same way as a *clique partitioning* problem. For registers, for example, all values in the DFG, which needed to be stored, were mapped to nodes in a graph, called *compatibility graph*. An edge between two nodes indicated that the nodes were *compatible*, which meant they could share the same resource. In order to find the minimum number of resources (registers, in this case) needed, a clique partitioning algorithm was applied to the graph to find the minimum number of cliques. Each clique represented a final register, which would be shared by all values/nodes in the clique. The same formulation was applied to functional units and busses.

Figure 7 presents an example of this approach. Figure 7a shows a scheduled DFG with six operations and five values that need storage. The compatibility graphs for functional units and registers are given in Figure 7b and 7c. Following the conditions for resource sharing given above, the algorithm derives the compatibility edges (for elements that can be shared) and insert them in the compatibility graphs. The next step of the algorithm is to find the minimum number of cliques in each graph, which are shown in dotted circles in Figure 7b and 7c. As a result, this

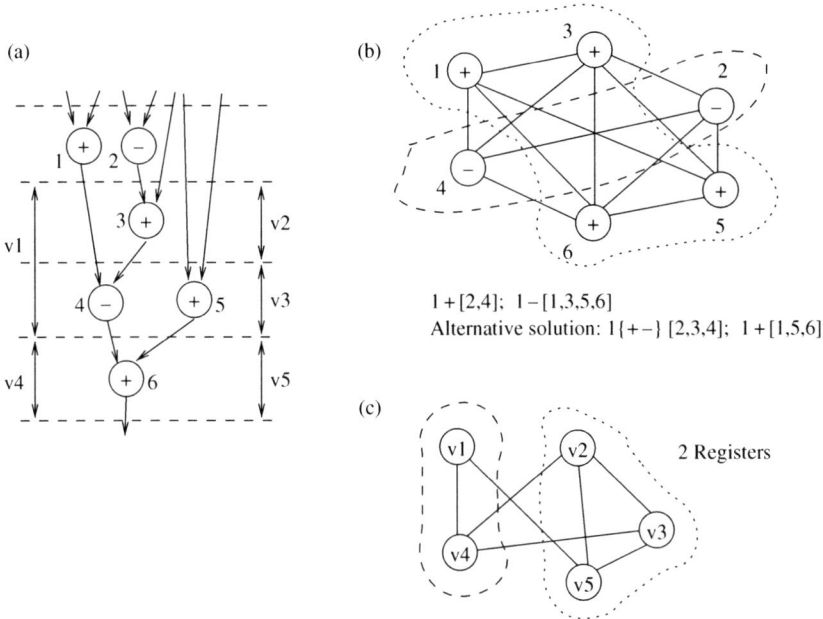

(a)

(b)

1 + [2,4]; 1 − [1,3,5,6]
Alternative solution: 1{+−} [2,3,4]; 1 + [1,5,6]

(c)

2 Registers

Figure 7. Resource sharing using compatibility graphs and clique partitioning

example needs only one adder, one subtracter and two registers for implementing all the operations and values in the DFG.

Although this result is optimal in number of cliques, there is no guarantee that the final logic will be optimal in area and/or delay because there are many factors involved in building the final hardware which are not fully considered. For example, this simple solution only looks at the final number of elements, but ignores completely the other hardware *stitching* elements, which are not explicit in the DFG, such as the multiplexers and control logic. Moreover, there may be more than one minimal clique cover for a graph, possibly requiring different functional units. In Figure 7b, another minimum clique solutions would result in one adder-subtracter unit, plus a separate adder.

Facet (Tseng and Siewiorek, 1996) recognized this problem and proposed an algorithm for finding a clique cover based on *weighed* edges. Each compatibility edge was assigned a weight based on the estimated final effect of merging the two elements. This estimation was purely local and often not very effective in improving the final result, especially when large numbers of cliques and nodes were present.

Although clique partitioning is an NP-hard problem, the numbers of nodes in compatibility graphs in realistic designs are not extremely large, which makes even exact algorithms applicable. Moreover, the graphs are reasonably sparse so that heuristic algorithms also give optimal or near-optimal results in most cases.

Bergamaschi et al., 1995 used a similar approach as in FACET, but using a formulation based on conflict graphs instead of compatibility. The approach relied

on a coloring algorithm (Philips, 1990) for finding the minimum number of colors
or elements. A variation of this approach applied to register sharing only was
introduced by Kurdahi and Parker, 1987 in the REAL system. They realized that
for scheduled DFG's, the register sharing problem can be modeled as the left-edge
problem (Hashimoto and Stevens, 1971) and solved in polynomial time.

More recently, Raje and Bergamaschi, 1997 proposed a new algorithm for performing interleaved register and functional unit sharing within a clique partitioning
formulation. They present algorithms for accurately estimating the effects of merging two elements, in area, required multiplexers and control logic, as well as the
effect on future mergings.

Branch-and-bound algorithms have also been proposed by Pangrle, 1988 and
Marwedel, 1986 for searching the allocation design space. This is global search,
which can be very expensive, and heuristics for limiting the search space have also
been developed.

3.7   Controller Generation

So far we have described techniques for generating a schedule and a data-path, but
not how these parts are put together into a working design – this is the task of the
controller. Controller generation involves two main parts:
1. the creation of the state machine and the state transition equations, and
2. the derivation of all control signals required for controlling data-path events.

These events involve, for example, loading a value into a register, selecting a
particular multiplexer input, selecting a function to be performed by a multi-function
unit, or driving a bus with one driver while the others go to high-impedance (tri-
state). For each event to take place one or more control signals have to be active at
the right time (that is, the right state and the right conditions).

Two basic controller architectures have been used by behavioral synthesis
systems:
1. Microprogrammed machines, and
2. Hardwired finite-state machines.

In microprogrammed machines, the values of the control signals for all control
steps are stored in memory, and there is a sequencer which iterates through the
addresses and retrieves the control signals for each control step (Rabaey et al.,
1988). The approach is simpler to generate but it can produce area-inefficient results
if the number of control signals and states becomes very large.

In hardwired finite-state machines, the states are encoded into registers and the
transition equations are implemented as random logic (Bergamaschi et al., 1995;
Knapp et al., 1995; Thomas and Fuhrman, 1988). The HIS system (Bergamaschi
et al., 1992) uses Binary-decision diagrams (BDDs) to generate the control equations of controller. During scheduling and allocation, HIS computes the transition
equations and the required control signals as BDDs, such that there is one BDD
for each register load-enable, each multiplexer control input, and so on. After
scheduling and allocation, the BDDs mapped to random logic.

4. PROS AND CONS OF BEHAVIORAL SYNTHESIS

Behavioral synthesis has definitely reached a stage where it can be used in a production environment. The quality of its results and the productivity advantages that it brings have warranted it a place in many design methodologies.

Among the commercial and academic synthesis systems which have reported use in an industrial environment are: IBM's HIS (Bergamaschi et al., 1995) Synopsys' Behavioral Compiler (Knapp et al., 1995), Alta/Cadence's Visual Architect, Menthor Graphics' Monet and Mistral, Amical (Berrebi et al., 1996), IMEC's Cathedral Compiler (Rabaey et al., 1988) and Philips' Phideo (Lippens et al., 1991).

Despite its commercial presence, there are many pending issues in behavioral synthesis which still prevent its widespread use. The most important issues, involving both algorithms and methodology, are listed below.

– **Change in cycle-by-cycle behavior**. Current digital design methodologies rely on cycle-accurate simulation. The change in cycle-by-cycle behavior introduced by scheduling creates a major challenge to verification, which has not yet been solved.

   In modern IC design methodologies, synthesis and verification work in tandem. For example, if a designer simulates the input specification in order to verify its functionally, then it must make sure that the synthesis (and layout) tools correctly implement this functionality. This is needed primarily because there are many different tools and manual intervention steps in the process of going from a language specification to a final layout, and it is important to have an independent tool for verifying the equivalence between what is simulated and what is finally going to be fabricated.

   Equivalence checking between a specification and an implementation implies that for all possible sequences of inputs, the values observed at the outputs match exactly, cycle-by-cycle. The task of scheduling in behavioral synthesis may change the cycle-by-cycle behavioral of the implementation by introducing extra cycles is order to reduce resources or minimize cycle time. As a result, the functionality that is simulated (the specification) may not match exactly what is implemented. This is significant drawback for the designer because it means that the simulation of the design would necessarily have to be repeated after synthesis, which would be very costly (Bergamaschi and Raje, 1997b).

   Different schemes have been proposed to alleviate this problem. Ku and De Micheli, 1992 and Ly et al., 1995, for example, developed scheduling algorithms which understand timing dependencies between events. Bergamaschi and Raje, 1997b proposed an approach for comparing simulation results before and after high-level synthesis through the concept of *observable time windows*.

   Possible solutions to this problem may involve different scheduling approaches, dedicated verification tools (which *understand* scheduling), and more integrated design and verification methodologies.

– **Quality of results**. Despite significant improvements to behavioral synthesis algorithms, the quality of results could still be improved. In general, synthesis tools are very dependent on the input description, and behavioral synthesis is no exception. Better results may come from more general algorithms for control and data-path, better handling of input constraints and don't care conditions, and more efficient ways to control the output of synthesis.

- **Run-time performance**. Behavioral synthesis is usually regarded as the first synthesis step, which converts a language description into netlist (RTL or gate-level) that can be processed by lower level tools. Because it is used a *front-end* synthesis tool, the level of run-time performance expected from a behavioral synthesis tool a usually one or two orders of magnitude faster than lower level tools (such as logic synthesis), almost approaching the performance of language compilers. Current algorithms need to be further improved in order to deliver such performance level.

- **Tuned to lower level tools**. In order to get good results out of behavioral synthesis it is important to consider also the lower level tools which are supposed to optimize and map the output RTL/gate-level netlist. The final evaluation needs to be done at the technology mapped netlist and layout, and not simply at measuring numbers of adders and multiplexers. In order to get better results from logic synthesis and layout it is important that the RTL/gate-level netlist produced by behavioral synthesis be *tuned* to those tools. For example, behavioral synthesis can identify parts of the design as control or data-path and pass that information to logic synthesis, which then can apply optimizations dedicated to control or data-path synthesis, and similarly for layout tools.

- **Methodology issues**. Integration of behavioral synthesis in the overall design methodology is fundamental. The adoption of behavioral synthesis may require changes or additions to other parts of the methodology, such as, verification (both simulation and formal), synthesis and layout. Unless all parts work together, the benefits of behavioral synthesis will not be fully achieved.

- **Designers education**. One should not underestimate the learning curve of a new technology. Designers today are still used to writing RTL and gate-level descriptions, and making the transition to sequential and behavioral specifications is not trivial. This is an area that will improve with time, both because of new students coming out of universities with synthesis training, as well as, productivity pressures that will force designers to change. Moreover, as mentioned above, synthesis is dependent on the style of the input description and in order to get good quality results, it is important to understand what synthesis does and how to control it.

The need for significant productivity jumps continues to increase, and despite some of its shortcomings, behavioral synthesis has shown that it can help designers achieve good quality results, with much higher productivity than RTL or gate-level synthesis. The systems that can best address the issues above will be the most successful in the market place.

REFERENCES

Aho, A., Sethi, R, and Ullman, J. (1996). Compilers, Principles, Techniques and Tools. Addison-Wesley, Reading, MA.

Bergamaschi, R. A. and Allerton, D. J. (1988). A graph-based silicon compiler for concurrent VLSI systems. In Proceedings of the IEEE CompEuro Conference, pages, 36-47, Brussels. IEEE.

Bergamaschi, R. A, Camposano, R., and Payer, M. (1991a). Area and performance optimizations in path-based scheduling. In Proceedings of the European Conference on Design Automation, Amsterdam, The Netherlands. IEEE.

Bergamaschi, R. A, Camposano, R., and Payer, M. (1991b). Scheduling under resource constraints and module assignment. INTEGRATION, the VLSI Journal, 12:1-19.

Bergamaschi, R. A, Camposano, R., and Payer, M. (1992). Allocation algorithms based on path analysis. INTEGRATION, the VLSI Journal, 13:283-299.

Bergamaschi, R. A, O'Conner, R., Stock, L., Moricz, M., Prakash, S., Kuehlmann, A., and Rao, D.S. (1995). High-Level synthesis in an industrial environment. IBM Journal and Research and Development, 39(1/2):131-148.

Bergamaschi, R. A., and Raje, S. (1997a). Control-flow versus data-flow-based scheduling: Combining both approaches in an adaptive scheduling system. IEEE Transactions on VLSI Systems, 5(1).

Bergamaschi, R. A., and Raje, S. (1997b). Observable time windows: Verifying high-level synthesis results. IEEE Design & Test of Computers, 14(2):40-50.

Berrebi, E., Kission, P. Vernalde, S., De Troch, S., Herluison, J.C., Frehel, J. Jerraya, A., and Bolsens, I. (1996). Combined control flow dominated and data flow dominated synthesis. In Proceedings of the 33rd ACM/IEEE Design Automation Conference, pages 573-576, Las Vegas, NV. ACM/IEEE.

Camposano, R. (1991). Path-based scheduling for synthesis. IEEE Transactions on Computer-Aided Design, CAD-10(1):85-93.

Camposano, R. and Rosentiel, W (1989). Synthesizing circuits from behavioral descriptions. IEEE Transactions on Computer-Aided Design, CAD-8(2):171-180.

Camposano, R. and Tabet, R.M. (1989). Design representation for the synthesis of behavioral VHDL models. In Proceedings 9th International Symposium on Computer Hardware Description Languages and Their Applications, pages 49–58, Washington, D.C. Elsevier Science Publishers B.V.

Cloutier, R.J. and Thomas, D.E. (1990). The combination of scheduling, allocation, and mapping in a single algorithm. In Proceedings of the 27th ACM/IEEE Design Automation Conference, pages 71-76. ACM/IEEE.

Davidson, S. Landskov, D. Shriver, B., and Mallet, P. (1981). Some experiments in local microcode compaction for horizontal machines. IEEE Transactions on Computers, C-30(7):460-477.

Gebotys, C.H. (1993). Throughput optimized architectural synthesis. IEEE Transactions on VLSI Systems, 1(3):254-261.

Goosens, G. Vanderwalle, J., and De Man, H. (1989). Loop optimization in register-transfer scheduling for dsp systems. In Proceedings of the 26th ACM/IEEE Design Automation Conference, pages 826-831. ACM/IEEE.

Hafer, L.J. and Parker, A. C. (1983). A formal method for the specification, analysis, and designs of register-transfer level digital logic. IEEE Transactions on Computer-Aided Design, CAD-2(1):4-18.

Hashimoto, A. and Stevens, J. (1971). Wire routing by optimizing channel assignment within large apertures. In Proceedings within large apertures. In Proceedings of the 8th Design Automation Workshop, pages 155-169.

Hitchcock III, C and Thomas, D. (1983). A method of automated data path synthesis. In Proceedings of the 20th ACM/IEEE Design Automation Conference, pages 484-489. ACM/IEEE.

Hwang, C.-T., Lee, J.-H., and Hsu, Y.-C. (1991). A formal approach to the scheduling problem in high-level synthesis. IEEE Transactions on Computer-Aided Design, CAD-10(4):464-475.

Jerraya, A., Park, I. O'Brien, K. (1993). AMICAL: An interactive high-level synthesis environment. In Proceedings of the European Conference on Design Automation, Paris. IEEE.

Knapp, D., Ly, T., MacMillen, D. and Miller, R. (1995). Behavioral synthesis methodology for hdl-based specification and variation. In Proceedings of the 32nd ACM/IEEE Design Automation Conference, pages 286-291. ACM/IEEE.

Ku, D. C. and De Micheli, G. (1992). Relative scheduling under timing constraints: Algorithms for high-level synthesis of digital circuits. IEEE transactions on Computer-Aided Design, CAD-11(6).

Kucukcakar, K. and Parker, A. (1990). Data path tradeoffs using MABAL. In Proceedings of the 27th ACM/IEEE Design Automation Conference, pages 511-516. ACM/IEEE.

Kuelhmann, A. and Bergamaschi, R. A. (1992). Timing analysis in high-level synthesis. In Proceedings of the IEEE International Conference on Computer-Aided Design, pages 349-354. IEEE.

Kurdahi, F. J. and Parker, A. C. (1987). REAL: A program for register allocation. In Proceedings of the 24th ACM/IEEE Design Automation Conference, pages 210-215. ACM/IEEE.

Lippens, P., van Meerbergen, J. L., va der Werf, A., and Verhaegh, W. F. J. (1991). PHIDEO: A silicon compiler for high speed algorithms. In Proceedings of the European Conference on Design Automation Amsterdam The Netherlands. IEEE.

Lis, J. and Gajski, D. (1988). Synthesis from VHDL. In Proceedings of the IEEE International Conference on Computer Design, pages 378-381. IEEE.

Ly, T., Knapp, D., Miller, R., and MacMillen, D. (1995). Scheduling using behavioral templates. In Proceedings of the 32nd ACM/IEEE Design Automation Conference, pages 101-106. ACM/IEEE.

Marwedel, P. (1986). A new synthesis algorithm for the Mimola software system. In Proceedings of the 23rd ACM/IEEE Design Automation Conference, pages 271-277. ACM/IEEE.

McFarland, M. C. (1978). The value Trace: A data base automated digital design. Technical Report DRC-01-4-80, Design Research Center, Carnegie-Mellon University.

McFarland, M. C. (1978). Using bottom-up design techniques in the synthesis of digital hardware from abstract behavioral descriptions. In Proceedings of the 23rd ACM/IEEE Design Automation Conference. ACM/IEEE.

Nestor, J. and Thomas, D. (1986). Behavioral Synthesis with interfaces. In Proceedings of the IEEE International Conference on Computer-Aided Design, pages 112-115. IEEE.

Pangrle, B. (1988). Splicer: A heuristic approach to connectivity binding. In Proceedings of the 25th ACM/IEEE Design Automation Conference. ACM/IEEE.

Pangrle, B. M. and Gajski, D. D. (1986). State synthesis and connectivity binding for microarchitecture compilation. In Proceedings of the IEEE International Conference on Computer-Aided Design, pages 210-213. ACM/IEEE.

Papadimitriou, C. H. and Steiglitz, K. (1982). Combinational Optimization: Algorithms and Complexity. Prentice-Hall, Englewood Cliffs, NJ.

Parker, A. C., Pizarro, J. T., and Mlinar, M. (1986). MAHA: A program for data-path synthesis. In Proceedings of the 23rd ACM/IEEE Design Automation Conference, pages 461-466. ACM/IEEE.

Passos, N. L., Sha, E.H.-M., and Bass, S.C. (1994). Loop pipelining for scheduling multi-dimensional systems via rotation. In Proceedings of the 31st ACM/IEEE Design Automation Conference, pages 485-490. ACM/IEEE.

Paulin, P. G. and Knight, J. P. (1989). Force-directed scheduling for the behavioral synthesis of ASIC's. IEEE Transaction on Computer-Aided Design, CAD-8(6):661-679.

Philips, T. K. (1990). New algorithms to color graphs and find maximum cliques. Technical Report Computer Science, RC-16326 (#72348), IBM Research Division, T. J. Watson Research Center, Yorktown Heights, NY 10598.

Rabaey, J., De Man, H. Vanhoof, J. and Catthoor, F. (1988). CATHEDRAL II: A synthesis system for multiprocessor dsp systems. In Gajski, D.D., editor, Silicon Compilation, pages 311-360. Addison-Wesley.

Raje, S. and Bergamaschi, R. A. (1997). Generalized resource sharing. In Proceedings of the IEEE International Conference on Computer-Aided Design, pages 326-332. San Jose. IEEE.

SRC (1994). Design Needs for the 21st Century: White Paper. Semiconductor Research Corporation.

Thomas, D. E. and Fuhrman, T. E. (1988). Industrial uses of the system architect's workbench. In Gajski, D. D., editor, Silicon Compilation, pages 307-329. Addison-Wesley.

Thomas, D. E., Lagnese, E. D., Walker, R. A., Nestor, J. A., Rajan, J. V., and Bergamaschi, R. L. (1990). Algorithm and Register-Transfer Level Synthesis: The system Architect's Workbench. Kluwer Academic Publishers, The Netherlands.

Tseng, C.J. and Siewiorek, D. P. (1996). Automated synthesis of data paths in digital systems. IEEE Transactions on Computer-Aided Design, CAD-5(3):379-395.

Verhaeg, W. F. J., Lippens, P. E. R., Aarts, E. H. L., Korst, J. H. M., van Meerbergen, J. L. and van der Werf, A. (1995). Improved force-directed scheduling in high-throughput digital signal processing. IEEE Transaction on Computer-Aided Design, CAD-14(8):945-960.

Wakabayashi, K. and Tanaka, H. (1992). Global scheduling independent of control dependencies based on condition vectors. In Proceedings of the 29th ACM/IEEE Design Automation Conference, pages 112-115. ACM/IEEE.

Wolf, W. Takach, A., Huang, C.-Y and Manno, R. (1992). The Princeton University behavioral synthesis system. In Proceedings of the 29th ACM/IEEE Design Automation Conference, pages 182-187. ACM/IEEE.

CHAPTER 7

HARDWARE/SOFTWARE CO-DESIGN

A. JERRAYA, J.M. DAVEAU, G. MARCHIORO, C. VALDERRAMA,
M. ROMDHANI, T. BEN ISMAIL, N.E. ZERGAINOH, F. HESSEL, P. COSTE,
PH. LE MARREC, A. BAGHDADI, AND L. GAUTHIER
*TIMA Laboratory, System-Level Synthesis Group, 46 Av. Felix Viallet, 38031 Grenoble Cedex,
FRANCE, +33 476 57 47 59, +33 476 47 38 14, Ahmed.Jerraya@imag.fr*

Abstract: Co-design is an important step during rapid system prototyping. Starting from a system-level specification, Co-design produces a heterogeneous architecture composed of software, hardware, and communication modules. This paper gives a taxonomy of co-design starting from a system-level specification and producing a heterogeneous architecture including the descriptions of hardware and software. Co-design is generally decomposed into four refinement steps: system-level specification, system-level partitioning, communication synthesis, and architecture generation. However, the co-design process depends on the kind of input language (synchronous/asynchronous, single thread/multi-thread) and the target architecture (mono-processor/multi-processor). The main co-design concepts are also detailed through the presentation of a co-design tool called COSMOS

Keywords: Hardware/Software Co-design, System Design Model, System-Level Synthesis

1. INTRODUCTION

The term co-design denotes a joint development of hardware and software components in order to obtain a complete system design. Co-design is not exactly the latest novelty in the field of system design. However, the joint specification, design and synthesis of mixed hardware/software systems is a recent issue.

The fields of design, specification and synthesis of mixed hardware/software systems are emerging and becoming more and more popular [1, 2, 3, 4]. The renewed interest in co-design is driven by the increasing complexity of the design and the need for early prototypes needed to validate the specification and to provide the customer with feedback during the design process [13].

Co-design is an important step in rapid system prototyping. Figure 1 shows a typical Rapid System Prototyping (RSP) design flow [13]. The design starts with

133

R. Reis et al. (eds.), Design of Systems on a Chip, 133–158.
© 2006 *Springer.*

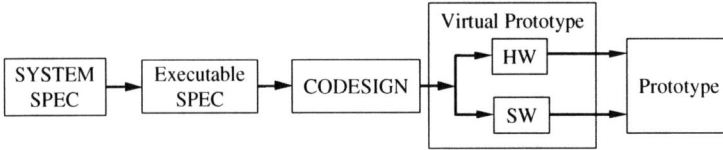

Figure 1. Co-design Environment

a system specification. This may be a non-executable specification given as a set of requirements. During the next step, an executable specification is produced. This may be given in an existing language such as SDL, StateChart or JAVA. At this stage, the specification is composed of a set of interacting modules. The next step is Co-design. It includes several design decisions and refinement steps. The goal of Co-design is to produce an abstract architecture composed of a set of communicating processors that implements the initial specification.

The Co-design process may be iterative. The main steps needed in order to transform a system-level specification into a mixed hardware/software one are system partitioning, communication synthesis and architecture generation. The output of architecture generation is a heterogeneous architecture composed of hardware and software modules. This step is decomposed into two tasks: virtual prototyping and architecture mapping (prototyping). Virtual prototyping produces a simulatable model of the system. The architecture mapping step produces an architecture that implements (or emulates) the initial specification.

2.   TAXONOMY OF CO-DESIGN

Several projects reported in the literature – SpecSyn at Irvine [46], CODES at Siemens [8], SDW at Italtel [28], Thomas approach at CMU [12], Gupta and DeMicheli approach at Stanford [2], Ptolemy at Berkeley [3, 4], RASSP [15] – are trying to integrate both the hardware and the software in the same design process. The key points of the state of the art may be summarized to three main concepts:
– the input specification,
– the considered target architecture,
– the co-design scheme.

2.1   System-Level Specification Models

The system-level specification of a mixed hardware/software application may follow one of two schemes [17]:
– Homogeneous specification: a single language is used for the specification of the overall system including both hardware and software parts.
– Heterogeneous modeling: specific languages are used for hardware parts and software parts, a typical example is the mixed C-VHDL model.

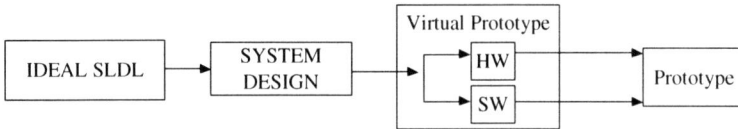

Figure 2. Homogeneous Modeling

Each of these two modeling strategies implies a different organization of the co-design environment.

Homogeneous modeling implies the use of a single specification language for the modeling of the overall system. A generic co-design environment based on homogeneous modeling is shown in Figure 2. Co-design starts with a global specification given in a single language. This specification may be independent of the future implementation and the partitioning of the system into hardware and software parts. In this case co-design includes a partitioning step aimed to split this initial model into hardware and software. The outcome is an architecture made of hardware processors and software processors. This is generally called "virtual prototype" and may be given in a single language or different languages (e.g. C for software and VHDL for hardware).

The key issue with such co-design environments is the correspondence between the concepts used in the initial specification and the concepts provided by the target model (virtual prototype). For instance the mapping of the system specification language including high-level concepts such as distributed control and abstract communication onto low-level languages such as C and VHDL is a task of major importance [18, 19].

Several co-design environments follow this scheme. In order to reduce the gap between the specification model and the virtual prototype, these tools start with a low-level specification model. Cosyma starts with a C-like model called C^x [20, 21]. VULCAN starts with another C-like language called hardwareC [22]. Lycos [43] starts with C. Several co-design tools start with VHDL [23]. Only few tools tried to start from a high-level model. These include Polis [24] that starts with an Esterel model [25, 26], SpecSyn [27, 28] that starts from SpecChart [29, 30, 31] and [19] that starts from LOTOS [32]. COSMOS is a co-design tool that starts from SDL, it will be detailed later.

Heterogeneous modeling allows the use of specific languages for the hardware and software parts. A generic co-design environment based on a heterogeneous model is given in Figure 3. Co-design starts with a virtual prototype when the hardware/software partitioning is already done. In this case, co-design is a simple mapping of the software and the hardware parts on dedicated processors.

The key issues with such a scheme are validation and interfacing. The use of a multi-language specification requires new validation techniques able to handle a multi-paradigm model. Instead of simulation we will need co-simulation and instead of verification we will need co-verification. Additionally, multi-language specification brings up the issue of interfacing subsystems which are described in

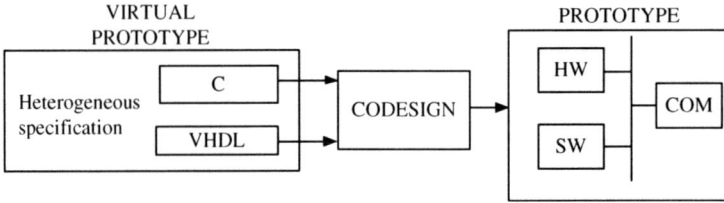

Figure 3. Heterogeneous Modeling

different languages. These interfaces need to be refined when the initial specification is mapped onto a prototype.

Coware [33] and Seamless [34] are typical environments supporting such a co-design scheme. They start from a mixed description given in VHDL or VERILOG for hardware and C for software. All of them allow for co-simulation. However, only few of these system allow for interface synthesis [35].

2.2   Towards System-Level Multilanguage Modeling and Co-simulation

Most of the existing system specification languages are based on a single paradigm. Each of these languages is more efficient for a given application domain. For instance some of these languages are more adapted to the specification of state-based specifications (SDL or StateChart), some others are more suited for data flow and continuous computation (LUSTRE, Matlab), while many others are more suitable for algorithmic description (C, C++).

When a large system has to be designed by separate groups, they may have different cultures and expertise with different modeling styles. The specification of such a large design may lead each group to use a different language which is more suitable for the specification of the subsystem they are designing according to its application domain and to their culture.

Figure 4 shows a generic flow for co-design starting from a multi-level specification. Each of the subsystems of the initial specification may need to be decomposed into hardware and software parts. Moreover, we may need to compose some of these subsystems in order to perform global hardware/software subsystems. In other words, partitioning may be local to a given subsystem or global to several subsystems. The

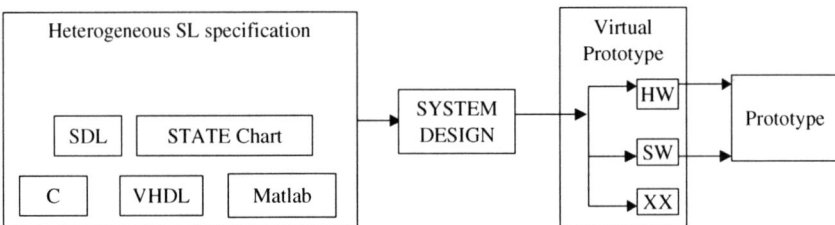

Figure 4. System-Level Heterogeneous Modeling

co-design process also needs to tackle the refinement of interfaces and communication between subsystems. In addition to hardware and software, a system may include other kinds of subsystems (XX in Figure 4) such as mechanical, analog devices etc.

As in the case of the heterogeneous modeling for system architecture, the problems of interfacing and multi-language validation need to be solved. In addition, this model brings up another difficult issue: language composition. In fact, when global partitioning is needed, the different subsystems need to be mapped onto a homogeneous model in order to be decomposed. This operation would need a composition format able to accommodate the concepts used for the specification of the different subsystems and their interconnect.

Only few systems in the literature allow such co-design models. These include RAPID [36] and the work described in [37]. Both systems provide a composition format able to accommodate several specification languages.

2.3   Target Architecture

The target architecture used to implement a design can be classified in one of the following cases:
- A mono-processor architecture composed of a main processor (acting as a top controller) and a set of hardware components (ASICs, FPGAs) [2, 8, 12],
- A distributed architecture [3, 4] composed of a set of processors running in parallel. Several configurations of processors (monoprocessor, multiprocessor) and several communication models (shared memory, message passing) may be used.

This classification is independent of the implementation technology. Both architectures may be implemented using a single chip, a multi-chip, a multi-board or even multi-machines.

The monoprocessor architecture (Figure 5a) corresponds to the classical datapath controller model. It represents a large class of applications ranging from simple

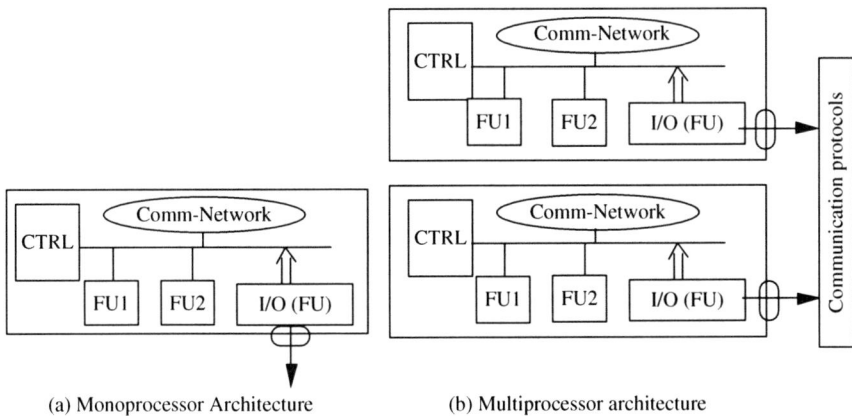

(a) Monoprocessor Architecture   (b) Multiprocessor architecture

Figure 5. Architectural model

circuits where the controller is a simple FSM, large systems where the controller is a programmable processor and the datapath includes large co-processors. Most co-design tools published in the literature are based on this model. A typical co-design architecture is composed of a top controller made of a programmable processor and a datapath that includes hardware accelerators. For instance, COSYMA [20, 21], VULCAN [22] and Lycos are based on a monoprocessor architecture.

2.4   Co-design Tools

Most current research issues on co-design fall into one of three categories:
1. ASIP (Application Specific Integrated Processor) co-design: In this case, the designer starts with an application, builds a specific programmable processor and translates the application into software code executable by the specific processor [38] [39] [40]. In this scheme the hardware/software partitioning includes the instruction set design [41]. In this case the cost function is generally related to area, execution speed and/or power.
2. Hardware/software synchronous system co-design: In this case the target architecture of co-design is a software processor acting as a master controller, and a set of hardware accelerators acting as co-processors. Within this scheme two kinds of partitioning have been developed: software-oriented partitioning [42] and hardware-oriented partitioning [43]. Most of the published works on co-design fall into this scheme. They generally use a simple cost function related to area, to processor cost for software and to speed for hardware. Vulcan [43], Codes [18], Tosca [45], and Cosyma [42] are typical co-design tools for synchronous systems.
3. Hardware/Software for distributed systems: In this case, co-design is the mapping of a set of communicating processes (task graph) onto a set of interconnected processors (processor graphs). This co-design scheme includes behavioral decomposition, processor allocation and communication synthesis [44] [46]. Most of the existing partitioning methods restrict the cost function to parameters such as real-time constraints [47] [48] or cost [49]. Coware [48] handles very well multiprocessor co-design during the latest design phases. However, it starts from a C/VHDL specification where partitioning is already done. SpecSyn [50] is a precursor for the co-design of multiprocessors. It allows automatic partitioning and design space exploration. However, SpecSyn does not help the designer to understand the produced architecture and it does not allow a co-design with partial solutions. Siera [51] provides a powerful scheme for co-design of multiprocessors based on the re-use of components. It allows co-design with partial solutions. However it does not provide automatic partitioning. Ptolemy provides a powerful environment for co-design of multiprocessors [52]. However, its partitioning is restricted to DSP applications and does not allow partial solutions [53]. Wolf's group studied system-level analysis/performance and co-synthesis of distributed multiprocessor architectures [48]. They developed an efficient algorithm to bound the computation time of a set of processes on a multiprocessor system, a synthesis method which takes into account communication and computation cost, and scheduling and allocation algorithms for multiprocessors.

The rest of the paper describes the anatomy of COSMOS, a co-design environment dealing with both multi-thread descriptions and multi-processor architectures. The next section gives an overview of COSMOS. Section 4 details the design models used during the co-design process. The design steps are introduced in Section 5. Finally, Section 6 discusses a co-design case through a design example using the COSMOS methodology. The goal is to present the status of the COSMOS environment and to outline future directions.

3.   COSMOS: A CO-DESIGN ENVIRONMENT

This section concentrates on the design flow within the COSMOS environment. Currently, COSMOS starts from the system-level specification language SDL, and produces a heterogeneous architecture including hardware descriptions in VHDL and software descriptions in C. As shown in Figure 3, the design process in the COSMOS environment is decomposed into four refinement steps: system-level specification, system-level partitioning, communication synthesis (including channel binding and channel mapping), and architecture generation (including virtual prototyping and architecture mapping). In this chart, the boxes are to be interpreted as activities, whereas the small circles correspond to intermediate models and libraries. These are expanded and represented as system-level graphs. These models and libraries are explained in the next sections. The rest of this section gives a brief overview of these refinement steps and design models. A comprehensive description of these steps will be given later (Section 5).

All the intermediate models (between SDL and C/VHDL) are represented in a design representation called SOLAR. SOLAR is an intermediate form for system-level concepts. SOLAR allows several levels of description.

In the present version of COSMOS, the design flow starts from an SDL description. However, all the subsequent synthesis steps make use of an intermediate form called SOLAR. The first step in COSMOS translates the initial description (SDL) into SOLAR. At this level, a system is represented as a set of communicating processes. The intermediate model which is produced, in the example of Figure 6, is composed of two processes (DU1, DU2) communicating through a channel unit (CU1). The unit DU2 is itself composed of two concurrent processes (ST1, ST2). The next step in the design flow is partitioning. The goal is to distribute the previous model into a set of communicating modules that will be mapped on separate processors in the final prototype. This step makes use of a library of communication protocols and produces a refined model composed of a set of communicating and heterogeneous processes organized in a graph where the nodes may be either processors or communication units. This model needs to be refined in order to fix the communication models. This step is called communication synthesis. It is composed of two operations. The first, called channel binding, selects an implementation protocol for each communication unit. The second, called channel mapping, distributes the protocol among the processors and specific communication controllers. The

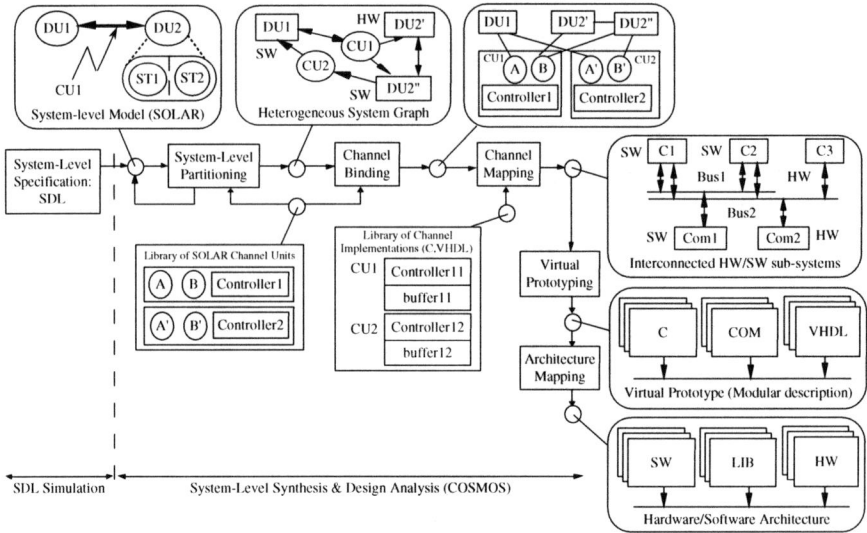

Figure 6. View of the COSMOS design flow

result is a set of interconnected processes communicating via signals and having the control of the communication distributed among the processes.

The final step is the architecture generation step. It starts with a set of interconnected subsystems (output of communication synthesis) and makes use of two methods in order to produce a prototype. The first produces a virtual prototype that can be used for simulation. The virtual prototype is an abstract architecture represented by VHDL for the hardware elements and by C for the software. This description is finally mapped on a multiprocessor architecture. The virtual prototyping step makes use of a task-level scheduling algorithm in order to produce the C code. Such techniques are needed when parallel processes have to be mapped on a single architecture.

Of course the design process will include lots of feedback loops in order to redesign parts or the full system. At any stage of the design, the user can cancel one or several design steps in order to explore new choices.

4. DESIGN MODELS FOR CO-DESIGN

As stated above the two main models used in a co-design environment are the input description and the target architecture. However the co-design process produces several intermediate models related to the different refinement steps. In COSMOS, all the co-design steps operate on a unified intermediate form called SOLAR [5]. This format is designed to meet two objectives:

1. Accommodate system level synthesis algorithms: The goal is to make easier the automation of the design steps. In fact SOLAR is based on an extended FSM model which is easy to operate on.

2. Accommodate several description languages: Although the present version starts from an SDL description, the long-term goal is to be able to use other existing languages such as, StateCharts, or JAVA. Our philosophy is to allow the designer to use one or more of these languages and to translate these descriptions into SOLAR.

The other important model in a co-design environment is the target architecture. COSMOS makes use of a modular and flexible multi-processor architecture.

This section details the main design models used by COSMOS. The intermediate form SOLAR will be introduced first, then the target architecture will be explained. It will be shown in the next section that the combination of these two models allows for modular design and design re-use of existing subsystems. This section outlines also the C-VHDL model produced by COSMOS.

4.1   SOLAR: A System-Level Model for Co-design

SOLAR is a design representation for system-level concepts. SOLAR allows several levels of description, starting from the level of communicating systems which contains a hierarchical structure of processes communicating via channels right down to the register transfer level and basic FSM descriptions. In addition, the communication schemes can be described separately from the rest of the system, this allows for modular specification and design. SOLAR's basic model is an extended FSM that allows the representation of hierarchy and parallelism. The main concepts are shown in Figure 7.

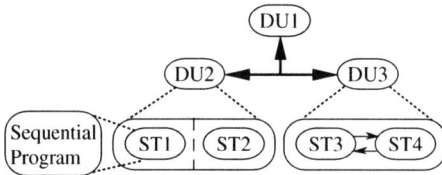

(a) Design hierarchy: DesignUnits, StateTables and ChannelUnits

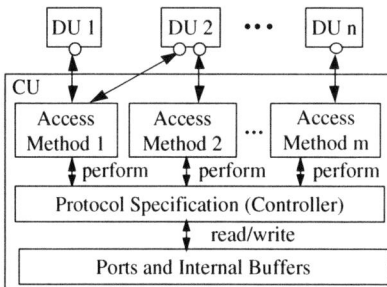

(b) SOLAR's communication model: the ChannelUnit

Figure 7. SOLAR: Basic Model

In SOLAR, a system is structured in terms of communicating DesignUnits (DUs).

A DesignUnit can either contain a set of other DesignUnits and communication operators known as ChannelUnits (CUs), or a set of transition tables modeled by the StateTable (ST) operator. This operator is used to model process-level hierarchy. StateTables can be executed in parallel (as indicated by the dotted line between ST1 and ST2 in Figure 4a) or serially. They can contain other StateTables, simple leaf states, state transitions and exceptions.

Communication between subsystems (or DesignUnits) is performed using SOLAR's ChannelUnit. It is possible to model most system-level communication properties such as message passing, shared resources and other more complex protocols. The ChannelUnit combines the principles of monitors and message passing. The model is known as the Remote Procedure Call or RPC. The RPC model is a mechanism that allows processes to communicate across message-carrying networks. The networking services are transparent to the user and communication is invoked using the semantics of a standard procedure call.

Figure 4b shows the basic configuration of the ChannelUnit. The ChannelUnit allows communication between any number of DesignUnits (a DesignUnit may be viewed as a system-level process). Access to the ChannelUnit is controlled by a fixed set of procedures known as methods (or services). These methods correspond to the visible part of the channel. In order to communicate, a DesignUnit needs to access at least one method. It achieves this through the use of a special procedure call statement known as a CUCall. In other words, the channel acts as a co-processor for the processes using it. The rest of the ChannelUnit is completely transparent to the user and consists of a set of ports linking the methods' parameters to the channel's controller. The controller guards the current state of the channel as well as conflict-resolution functions. The methods interact with the controller which in turn modifies the channel's global state and synchronizes the channel.

During the synthesis process, COSMOS uses an external library of CUs. A CU corresponds to either a standard protocol or a customized protocol described by the user. During partitioning and communication synthesis an abstract model of the channel is used. At the architecture generation step, an implementation of the channel is needed. This implementation may be the result of an early synthesis step using COSMOS or another design method. It may also correspond to an existing architecture.

Not only does this model enable the user to describe a wide range of communication schemes, it also separates the communication from the rest of the design, thereby allowing the re-use of existing communication schemes.

In order to illustrate this communication model and to give a flavor of SOLAR, a small example is given. Figure 8 gives more details about the concept and the use of a channel unit.

Figure 8a shows a conceptual view of a channel linking two processes (host and server). Figure 8b gives a brief outline of the SOLAR description of this system. The first DesignUnit, Host-Server is a structural view of the system. The two constituent DesignUnits, Host and Server, are instantiated and a netlist of the

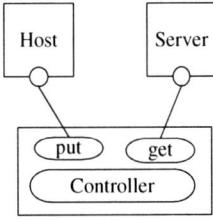

(a)

```
(SOLAR Host-Server
 (Design Unit Host-Server
  (Interface...)
  (Contents
   (Instance Host...)
   (Instance Server...)
   (Net (Joined...)...)
  )
 )
 (DesignUnit Host
  ...
   (CUCall put local_channel_name)
  ...
 )
 (DesignUnit Server
  ...
 ))
```

(b)

```
(Channelunit request_manager
 (view hl_spec
 (Interface

 (port reqin (direction in) (bit ))
 (port reqack (direction out) (bit ))
 (port datain (direction in) (integer))
 (port restart (direction in) (bit))
 (port rdy (direction in) (bit))
 (port b_full (direction out) (bit))
 (port inquire (direction out) (bit))
 (port inqack (direction in) (bit))
 (port dataout (direction out) (integer))

 (Method put
  (parameter request (integer))
  (statetable put
  (statelist put)
  (state put
   (if (= b_full '1')
  (then (wait (until (= b_full '0'))))
  )
 (assign datain request)
 (assign reqin '1')
 (wait (until (= reqack '1')))
 (assign reqin '0'))))

 (Method get
  (parameter request (integer))
  (statetable request
  (statelist req)
  (state req
   (wait (until (= inquire '1')))
   (assign rdy '1')
   (wait (until (= inqack '1')))
    (assign request dataout)
   (assign rdy '0')))))
```

(c)

```
(Contents
 (constant buf_size (integer))
 (variable (array buffer buf_size) (integer))
 (variable buffin (integer 1))
 (variable buffout (integer 1))
 (statetable request_m
 (statelist init send_recv)
 (entrystate init)
 (state init
  (assign buffin '1') (assign buffout '1')
  (nextstate send_recv))
 (state send_recv
  (paraction
  (globalaction (if (= restart '1') (then (nextstate init))))
 (statetable recv
 (statelist wait_req buf_full)
 (state wait_req
  (wait (until (= reqin '1')))
  (assign (member buffer buffin) datain)
  (assign reqack '1')
  (if (= (+1 (mod buffin buf_size)) buffout)
   (then (nextstate buf_full))
   (else (assign buffin (+1 (mod buffin buf_size)))
   (nextstate wait_req))))
 (state buf_full
  (assign b_full '1')
  (wait (until (/= (+1 (mod buffin buf_size)) buffout)))
  (assign b_full '0')
  (nextstate wait_req)))
 (state send
  (assign inquire '0')
  (if (= buffin buffout)
   (then (wait (until (/= buffin buffout)))))
  (assign inquire '1')
  (wait (until (= rdy '1')))
  (assign dataout (member buffer buffout))
  (assign inqack '1')
  (assign buffout (+1 (mod buffout buf_size)))
  (nextstate send)))))))))
```

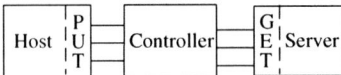

(d)

Figure 8. Different abstraction levels of a channel unit linking two processors: (a) Conceptual view, the only visible part of the channel are the communication primitives, (b) Extract of the system specification in SOLAR, the channel unit call, (c) Channel unit description in SOLAR, (d) channel unit implementation, the communication protocol is distributed over the communicating processors and a controller

connections between these two DesignUnits is given. This netlist will contain the name-mapping information for the channel that joins the two DesignUnits. The other units are representations of the functionality of the three constituent elements (the DesignUnits may themselves be structural). Two DesignUnits are used to model the host and server processes.

This description is independent of the communication protocol. The only information necessary is the fact that we can execute two Methods named get and put. With SOLAR, we may have several protocols executing the same methods. As we will see later, the channel binding is in charge of allocating channel units for the execution of these methods.

Figure 8c gives an extract of a SOLAR description of a channel unit. This description is composed of an interface and a contents. The interface describes the ports of the channel unit controller, and the methods describe the protocol of using these ports. In this case, the channel can be accessed through two methods put and get. This description may correspond to the abstraction of one or several existing communication units. The model allows to hide the implementation details of the channel. During the early stage of the design process, only the methods executed by a given channel are visible. The implementation of the channel is needed only during the channel mapping step. We will see later (Section 5) that this step replaces the abstract channels by an implementation. The result of this step is a set of interconnected modules. The communication modules are expanded in the caller process to control the exchange of data with the communication controller (see Figure 8d).

4.2   Hardware/Software Architectural Model

COSMOS uses a modular and flexible architectural model. The general model is given in Figure 9a, it is composed of three kinds of components: (1) Software components, (2) Hardware components, and (3) Communication components. This model serves as a platform onto which a mixed hardware/software system is mapped.

Communication modules come from a channel unit library, they correspond to existing communication models that may be as simple as a handshake or as complex as a layered network. For example, a communication controller may correspond to an existing interface circuit, an ASIC or some micro-code executing on a dedicated microprocessor.

The proposed architectural model is general enough to represent a large class of existing hardware/software platforms. It allows different implementations of mixed hardware/software systems. A typical architecture will be composed of hardware modules, software modules, and communication modules linked with buses. Figure 9b shows an example of a typical architecture supported by COSMOS.

4.3   Virtual Prototyping Using C-VHDL Models

The generated architecture, also called virtual prototype, is a heterogeneous architecture composed of a set of distributed modules, represented in VHDL for hardware elements and in C for software elements, communicating through communication

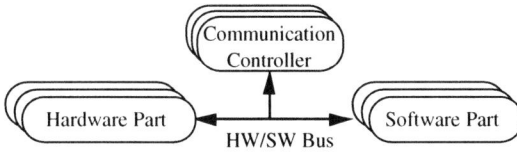

(a) Distributed Architecture

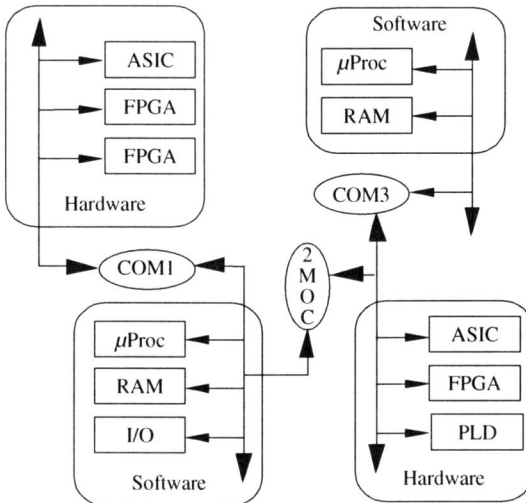

(b) Example of an Architecture Supported by COSMOS

Figure 9. Architectural models

modules (located into a library of components). The virtual prototype is a simulatable model of the systems. This model will be used for the production of the final prototype.

The prototyping (also called architecture mapping) produces an architecture that implements or emulates the initial specification. This step concentrates on the use of a virtual prototype for both co-synthesis (mapping hardware and software modules onto an architectural platform) and co-simulation (that is the joint simulation of hardware and software components) into a unified environment [54].

The joint environment for co-synthesis and co-simulation provides support for multiple platforms aimed at co-simulation and co-synthesis, communication between C and VHDL modules and coherence between the results of co-simulation and co-synthesis. The model used by the environment allows to separate module behavior and communication. The interaction between the modules is abstracted using communication primitives that hide implementation details of the communication unit. More details on this model are given in [54].

5.   DESIGN STEPS FOR CO-DESIGN

As stated above, the synthesis steps depend on the input description and the selected target architecture. COSMOS starts with a multi-thread description and produces a multi-processor architecture. The main steps needed are system specification, system partitioning, communication synthesis and architecture generation. These steps are detailed in the rest of this section.

5.1   Design Model Generation

The design flow starts from an SDL description. However, all the synthesis steps make use of SOLAR. SDL is also based on an extended FSM model. The main concepts handled by SDL are the system, the block, the process and the channel. The translation of most SDL concepts into SOLAR, with the exception of communication concepts, is fairly straightforward.

The translation of the SDL communication concepts (Channel and SignalRoute) leads to a re-organization of the description. In SDL, communication is based on message passing. Processes communicate through SignalRoutes. Channels are used to group all the communications between blocks. The communication methodology used by SOLAR allows such systems to be modeled. This scheme is summarized in Figure 10. Figure 10a shows a system example. It is composed of two communicating blocks. Block b1 contains two processes (P11, P12) that communicate with P21, a process that belongs to block b2. The translation of this system into SOLAR will produce the structure shown in Figure 10b. Each process is translated into a DesignUnit composed of a hierarchical FSM and a ChannelUnit.

Each SDL process is translated into a design unit composed of an extended FSM and a channel unit (see Figure 10). The SDL to SOLAR translation is performed automatically.

In this case, the SDL communication scheme was modeled through a SOLAR channel construct that explicitly describes the SDL communication scheme. In SDL, FIFOs are implicitly included in each process, and all signals, no matter what their type, are automatically stored in this FIFO. In addition, the ordering of

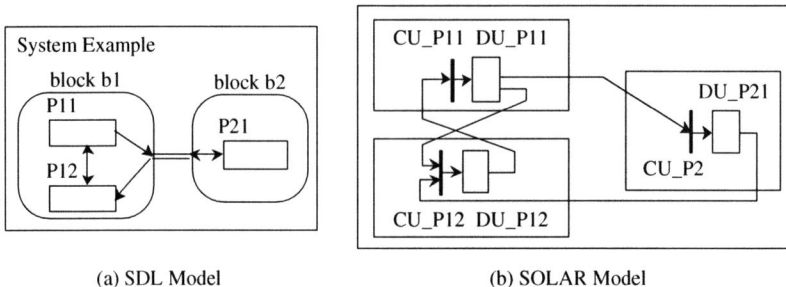

(a) SDL Model   (b) SOLAR Model

Figure 10. SDL model and corresponding SOLAR model

messages in the queues can be explicitly controlled. This type of communication scheme is not very efficient in terms of hardware. Thus, because of the high cost of communication, it would be more efficient, from a hardware point of view, to have larger initial SDL processes and less inter-process communication. This initial model will be improved during the communication synthesis step (see Section 5.3).

5.2 Hardware/Software Partitioning

Partitioning can be seen as a mapping of functional subsystems onto abstract processors. During this step a behavior may be distributed among several abstract processors. An abstract processor may also include several behaviors. A partitioning system may be either automatic or interactive. Figure 11 shows a typical partitioning process. The partitioning starts with two inputs: a system specification and a library of communication models. The output is a new model composed of a set of processors and a set of communication units. The library may be restricted to predefined models or extendible with user-defined models.

Partitioning maps a set of behaviors onto a set of processors. Starting from two communicating behaviors (DU1,DU2) the partitioning produced three design units and two channels. DU2 is split into two processes DU2′ and DU2″. Each of the design units in the new model will correspond to a processor in the final description. This step fixes the number of processors.

The partitioning step transforms a hyper-graph where the edges represent behavior and the vertices represent abstract communication channels into a hyper-graph where the edges correspond to either abstract processors (hardware or software) or abstract communication channels and the vertices represent physical links. An abstract processor corresponds to part of the design that will be implemented on a

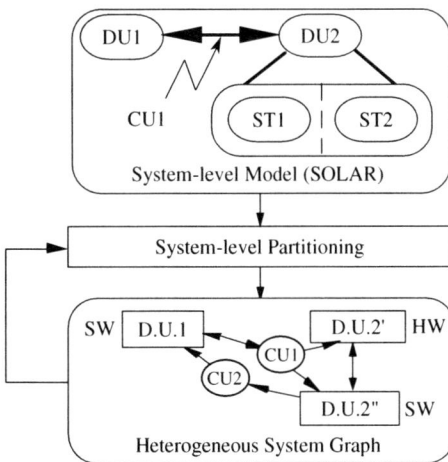

Figure 11. System-Level Partitioning

physical processor. The target may be a hardware processor or a software processor. An abstract processor communicates with its environment through communication primitives executed by channels (see Section 3.1). The result of system-level partitioning is a set of communicating and heterogeneous processors organized in a graph where the nodes may be either design units or channel units and the vertices of the graph may be signals or channel accesses. The partitioning step also determines which technology will be used for the implementation of each design unit. For example, a design unit may be implemented in pure hardware, in software running on an operating system or in micro-code adapted for a standard microprocessor. The choice is based on criteria such as execution time, rate of use, reprogrammability, re-use of existing components and technology limitations.

In COSMOS this step is achieved by the partitioning tool box called PARTIF [6]. PARTIF starts with a set of communicating processes organized in a hierarchical manner and described in SOLAR. Each process represents an extended FSM. Another input to PARTIF is a library of SOLAR communication models. The result of system-level partitioning is a set of communicating and heterogeneous processes organized in a graph where the nodes may be either design units or channel units and the edges of the graph may be signals or channel accesses.

PARTIF allows an interactive partitioning by means of five system-level transformation primitives. The three first primitives MOVE, MERGE, and CLUSTER allow the reordering of processes hierarchy and merging processes together to form a single process. The two second primitives are SPLIT and CUT. These allow splitting up one design unit to form inter-dependent design units for distribution purposes. PARTIF provides five system-level transformation primitives. A brief description of these primitives is given below:

- MOVE: This primitive allows to reorder the hierarchy (described into a DU) by transferring a process (a state or a StateTable) from one point to another.
- MERGE: The Merge primitive fuses two sequential processes (StateTables) into a single process. The objective may be to allow resource sharing among exclusive operations. In most cases merging processes implies a better use of resources. This means a reduction in the subsequent hardware required to implement the design.
- SPLIT: This primitive transforms a sequential machine into a parallel machine. This is achieved through the introduction of idle states and extra control signals. These will be used to control the activation and deactivation of the resulting FSMs. One can note that the value of the control signals gives the global state of the system.
- CUT: The primitive Cut transforms a set of parallel processes (StateTables) into a set of interconnected processes (DesignUnits). Parallel processes that share variables will be interconnected through communication channels that contain protocols governing access to these variables. A communication channel is assigned to each shared variable. The protocol of this channel depends on the type of access to the shared variable. This primitive defines interfaces between subsystems and also the communication protocols needed.

– CLUSTER: This primitive allows to cluster a set of processes (DesignUnits) into a single process. The primitive Cluster performs the reverse function of the primitive Cut. The objective is to obtain a set of behaviors to be mapped on a single abstract processor.

The sequencing of these primitives is decided by the user. The choice of which technology will be used for the implementation of each design unit is also decided by the user.

5.3   Communication Synthesis

The objective of communication synthesis is to transform a system containing a set of processes communicating via high-level primitives through channels into a set of interconnected processes communicating via signals and having the control of this communication distributed among the processes. This activity is decomposed into two tasks: channel binding and channel mapping [7, 18].

The channel binding algorithm is assumed to choose the appropriate set of channel units from the library of communication models to carry out the desired communication (see Figure 12). The main function of this step is to assign a communication unit for each communication primitive.

A channel unit, taken from this library, is selected in order to provide the desired services required by the communicating design units. This is similar to the binding/allocation of functional units in classic high-level synthesis tools. The communication between the subsystems may be executed by one of the schemes (synchronous, asynchronous, serial, parallel, etc.) described in the library. The choice of a given channel unit will not only depend on the communication to be executed but also on the performances required and the implementation technology

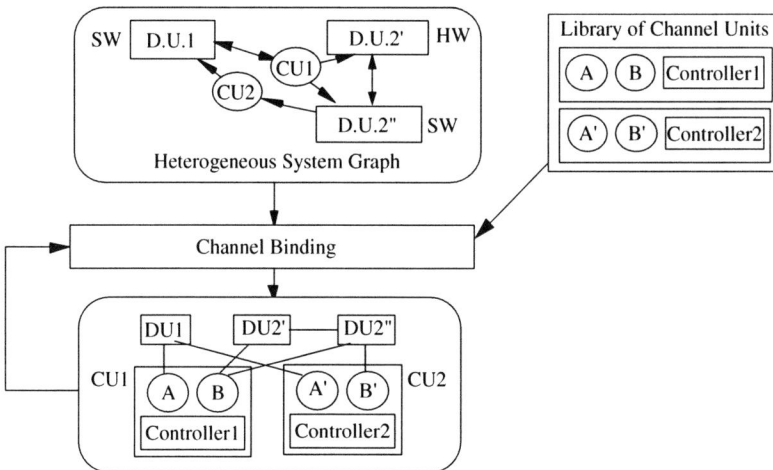

Figure 12. Channel Binding

of the communicating design units. The result of channel binding is a hyper-graph where the edges are either abstract processors or channel unit instances and the vertices correspond to channel accesses.

In COSMOS, communication synthesis starts with two SOLAR descriptions: a system graph and a library of communication models. In the present version of COSMOS, this task is done manually.

Channel binding selects a channel unit from the library for the execution of each communication primitive of the description. The selected channel unit should be able to execute the primitives used by the corresponding communicating processors. This step fixes the number of types of protocols to be used for each communication unit.

The channel mapping task replaces all channel units by distributing all of the information contained therein among the design units and specific communication controllers. These controllers are selected from a library of channel implementations (see Figure 7). This step may be regarded as going from communication design to implementation. The result of this task is a set of interconnected units. A unit may be an abstract processor or a component selected from the library of channels (Figure 13).

In COSMOS, this step expands the use of abstract communication units. All processes that execute a channel unit call to a particular method, will have this call replaced by a call to a local procedure implementing that method. The variables and signals used by the methods become ports in the modified design unit. In other words, the channel accesses are expanded into bundles of signals (ports) according to the methods used by the design units.

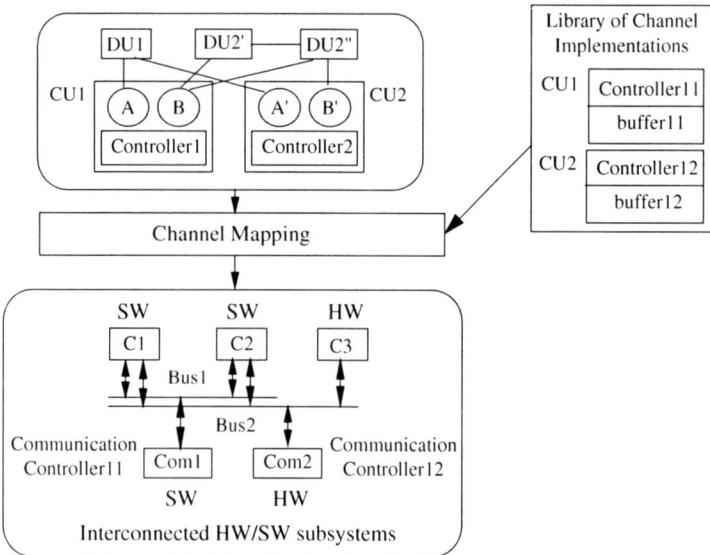

Figure 13. Channel Mapping

Channel mapping distributes the communication protocols over the processors and communication units. This step fixes the implementation of communication protocols and the interfaces of the subsystems.

A channel implementation is selected with regard to data transfer rates, memory buffering capacity, and number of control and data lines. All processes that execute a channel unit call to a particular method, will have this call replaced by a call to a local procedure implementing that method. The variables and signals used by the methods become ports in the modified design unit. In other words, the channel accesses are expanded into bundles of signals (ports) according to the methods used by the design units. The implementation of this local procedure call will depend on the implementation of the corresponding design unit. If the design unit is entirely software executing on a given operating system, method calls are expanded into system calls, making use of existing communication mechanisms within the system. If the design unit is to be executed on a standard micro-processor, the method becomes an access to a bus routine written in assembler. These two options are more software-oriented and require the user to partition his system into software and hardware elements before executing the communication synthesis. The design unit can also be executed as embedded software on a hardware datapath controlled by a micro-coded controller. In this case, the method call will become a call to a standard micro-code routine. Finally, the design unit may be implemented entirely in hardware.

5.4   Architecture Generation

Architecture generation starts with a set of interconnected hardware/software sub-systems (output of communication synthesis). As indicated above, this step is decomposed into two tasks: a virtual prototyping and an architecture mapping (prototyping). Virtual prototyping produces a simulatable model of the system. The architecture mapping step produces an architecture that implements (or emulates) the initial specification.

In COSMOS, virtual prototyping corresponds to a translation of SOLAR into executable code (VHDL and/or C). Each subsystem is translated independently. As shown in Figure 14, the output of virtual prototyping is a heterogeneous architecture represented by VHDL for the hardware elements (virtual hardware processors), C for the software elements (virtual software processors), and communication controllers (library components). In the case where several DUs need to be mapped onto the same software module (C-program) a scheduling step is performed during prototyping.

Each subsystem is mapped into a virtual processor: C for software, VHDL for hardware. The resulting virtual prototype allows a precise estimation of the resulting architecture.

The architecture mapping may be achieved using standard code generators to transform software parts into assembler code and hardware synthesis tools in order to translate hardware parts into ASICs or a virtual hardware processor (emulators).

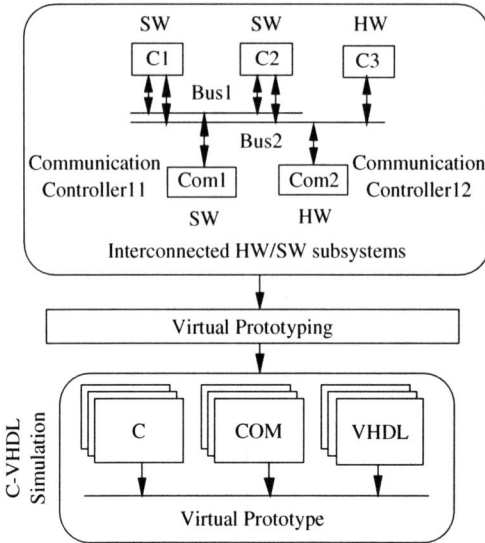

Figure 14. Virtual Prototyping

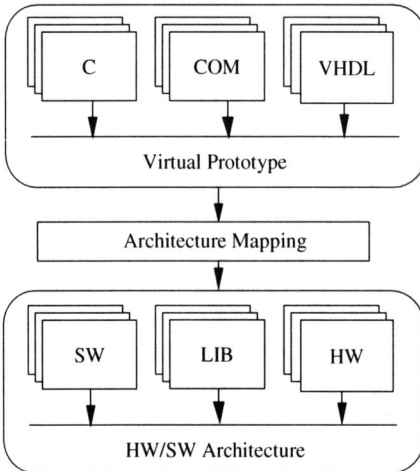

Figure 15. Architecture Mapping and Prototype Design

The result is an architecture composed of software, hardware, and communication components (Figure 15).

In COSMOS this step is achieved using standard compilers to transform C into assembler code for software parts and synthesis tools in order to translate the VHDL into hardware. At this level, an implementation of channels is needed. This implementation may be the result of an early synthesis step using COSMOS or another design method. It may also correspond to an existing architecture.

Prototyping is the final step of co-design. Each virtual processor is mapped onto a real (or prototype) processor. Software modules are compiled for existing micro-processors or specific cores, and hardware processors are mapped onto ASICs or FPGAs.

6. EVALUATION AND FUTURE DIRECTIONS

This section discusses a co-design case through a design example using the COSMOS methodology. The goal is to present the status of the COSMOS environment and to outline future directions.

Figure 16 illustrates the overall co-design process for a real-time system interface between a sensor producing digital signals that will be stored on a disk. The initial system is specified as a behavioral description acting between an acquisition unit and a storage unit (Figure 16a). This interface synchronizes the arrival of signals from the acquisition parts and their acceptance by the storage unit. The signals

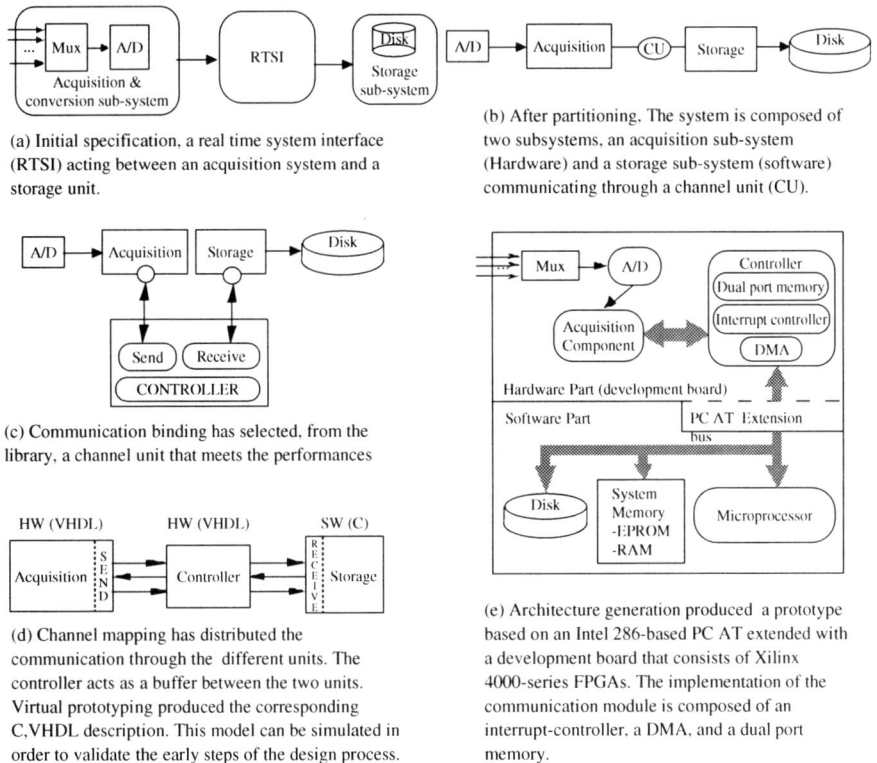

(a) Initial specification, a real time system interface (RTSI) acting between an acquisition system and a storage unit.

(b) After partitioning. The system is composed of two subsystems, an acquisition sub-system (Hardware) and a storage sub-system (software) communicating through a channel unit (CU).

(c) Communication binding has selected, from the library, a channel unit that meets the performances

(d) Channel mapping has distributed the communication through the different units. The controller acts as a buffer between the two units. Virtual prototyping produced the corresponding C,VHDL description. This model can be simulated in order to validate the early steps of the design process.

(e) Architecture generation produced a prototype based on an Intel 286-based PC AT extended with a development board that consists of Xilinx 4000-series FPGAs. The implementation of the communication module is composed of an interrupt-controller, a DMA, and a dual port memory.

Figure 16. The RTAS System Prototype

should be grouped into blocks before storage. A typical solution to such a design will be to use a dual port memory for this synchronization. The signals will be handled by a first processor that stores them in the memory using one of the ports. A second processor is used in order to transfer blocks of signals to the disk using the second port of the memory.

6.1   A Co-design Example

Starting from the initial specification, the system is first decomposed into an acquisition subsystem (Hardware) and a storage subsystem (software) communicating through a channel unit (Figure 16b). As explained above, this step is performed interactively through the application of a set of primitives under the control of the user. The next step is to select a communication protocol. The final parameters (size of the memory, blocking factors) of the communication units may be fixed later. This step links the communication primitives to a set of units that will execute them. This selection is performed during the communication binding step (Figure 16c). This step selected, from the library, a channel unit able to execute the different communication primitives. In COSMOS, the selection of the channel is performed by the user, however the SOLAR model is updated automatically. At this stage, several important functional decisions have already been made. All the subsequent steps are aimed to implement the different modules. However a crucial step is still ahead: reaching the real-time constraints. In the case of the use of a dual port memory, we need several evaluation steps in order to fix the size of the memory and the blocking factors. The main difficulty during this process is to estimate the speed (reaction time) of the two communication units. During the design of the two other processors we need to ensure that the resulting design meets the real-time constraints. The remaining design steps are channel mapping, virtual prototyping and architecture mapping. Channel mapping selects an implementation for the channel unit. This selection is also performed manually. The description is updated automatically. Virtual prototyping translates the resulting SOLAR into a C,VHDL description. In this case the communication unit is described in VHDL. This description models an existing scheme composed of an interrupt-controller, a DMA, and a dual port memory. The virtual prototype can be simulated in order to validate the early steps of the design process and in order to check some functional properties. The final step produces a prototype architecture (Figure 16e). In this case the prototype is based on an Intel 286-based PC AT extended with a development board that consists of Xilinx 4000-series FPGAs. During this step the VHDL description of the acquisition unit is mapped onto an FPGA and the C description corresponding to the storage unit is compiled for the 286.

6.2   Evaluation and Future Works

The previous sections presented COSMOS, an environment and a methodology for co-design. The methodology has already been proved through several designs. The most important design is a large robot application. It consists of the adaptive control of a robot arm with three fingers. This application controls an 18-engine system. It includes a mix of hardware and software. The environment provides already some of the methods described above. The SOLAR environment has already been well-tested. A first version of the partitioning tool has been developed [6]. In the present version, the design process is interactive: Most of the decisions are made by the designer, the system provides only primitives for automatic transformation of SOLAR descriptions. The sequencing of these primitives is made by the designer. The goal is to let the user do the intelligent work (e.g. decide which part will be in software and which part will be in hardware) and the system performs the fastidious and error-prone tasks.

The key issue in this scheme is the use of an abstract and general communication model. The separation between communication and computation (hardware and software) allows the re-use of existing communication models. Each communication unit can be specified at different levels of abstraction, for instance, functional (abstracted as a black box linking interconnected subsystems) and implementation views. The benefits from using external channel libraries at different levels of abstraction are the following:

- This scheme improves design re-use. For example, we can design a subsystem and then put it in the channel library in order to re-use it for building larger designs. A communication subsystem may also correspond to a design produced by external tools.
- A library of channels offers a wide range of different communication mechanisms and this helps the designer to select the appropriate communication protocol for his application. For instance, synchronous or asynchronous protocols, and blocking or non-blocking message exchange can be used.
- The different modules can be upgraded and re-designed independently. This modularity allows a large flexibility for maintaining and updating large designs.
- Since no restrictions are imposed to the communication models, this co-design process can be applied to a large class of applications.

In this paper only the design steps of a full co-design methodology have been discussed. Of course several other key issues are still to be worked out in order to have a complete design flow. These include:

1. Validation and verification of the initial specification. This is a classical problem for system design. Several tools exist for the validation and verification of SDL-like descriptions [55]. However, most of these tools act only at the functional level. Much work is still to be done in order to handle the real-time aspects.
2. Validation of the different design models against the initial specification: This validation is needed (even the tools are bug-free) in order to check the validity of the design decisions with regard to the constraints of the environment (real-time constraints).
3. Evaluation tools and models: as explained above, one of the crucial steps when deciding about the implementation of the different modules is to ensure that the resulting solution respects the real-time constraints. Only few works have already been done in this direction [20].

4. Automatic algorithms for partitioning and communication synthesis: These are needed in order to allow for fast and large exploration of the design space. The efficiency of such techniques will depend on the objective function which is used. This brings us again to the previous point: The need for good estimation methods.

5. Extending co-design to mechanical and other continuous parts of the design.

6. Combining virtual prototypes and prototypes: this would allow to validate a virtual prototype using an existing platform and to validate a prototype using a virtual model.

ACKNOWLEDGEMENT

This work was supported by France-Telecom/CNET under Grant 941B113, STMicroelectronics, Aerospatiale, PSA, ESPRIT programme under project COMITY 23015, ESPRIT/OMI programme under project CODAC 24129 and MEDEA programme under Projects SMT AT-403 and CIME AT-451.

REFERENCES

[1] G. Boriello, K. Buchenrieder, R. Camposano, E. Lee, R. Waxman, W. Wolf, "Hardware/Software Co-design", IEEE Design and Test of Computers, pp. 83-91, March 1993.

[2] R. Gupta, G. De Micheli, "Hardware-Software Cosynthesis for Digital Systems", IEEE Design and Test of Computers, pp. 29-41, September 1993.

[3] M. Chiodo, P. Giusto, A. Jurecska, L. Lavagno, H. Hsieh, A. Sangiovanni-Vincentelli, "A Formal Specification Model for Hardware/Software Co-design", Handouts of Int'l Wshp on Hardware-Software Co-design, Cambridge, Massachusetts, October 1993.

[4] A. Kalavade, E. A. Lee, "A Hardware-Software Co-design Methodology for DSP Applications", IEEE Design and Test of Computers, pp. 16-28, September 1993.

[5] A. A. Jerraya, K. O'Brien, "SOLAR: An Intermediate Format for System-Level Modeling and Synthesis", in "Computer Aided Software/Hardware Engineering", J. Rozenblit, K. Buchenrieder (eds), IEEE Press 1994.

[6] T. Ben Ismail, K. O'Brien, A. A. Jerraya, "Interactive System-Level Partitioning with PARTIF", Proc. EDAC'94, Paris, France, February 1994.

[7] K. O'Brien, T. Ben Ismail, A. A. Jerraya, "A Flexible Communication Modelling Paradigm For System-Level Synthesis", Handouts of Int'l Wshp on Hardware-Software Co-Design, Cambridge, Massachusetts, October 1993.

[8] K. Buchenrieder, A. Sedlmeier, C. Veith, "HW/SW Co-Design With PRAMs Using CODES", Proc. CHDL '93, Ottawa, Canada, April 1993.

[9] N. Dutt et al, "An Intermediate Presentation for Behavioral Synthesis", Proc. 27th DAC, pp. 14-19, Orlando, June 1990.

[10] W. Glunz, T. Kruse, T. Rossel, D. Monjau, "Integrating SDL and VHDL for System-Level Hardware Design", Proc. CHDL '93, Ottawa, Canada, April 1993.

[11] N.L. Rethman, P.A. Wilsey, "RAPID: A Tool for Hardware/Software Tradeoff Analysis", Proc. CHDL '93, Ottawa, Canada, April 1993.

[12] D.E. Thomas, J.K. Adams, H. Schmitt, "A Model and Methodology for Hardware-Software Co-design", IEEE Design and Test of Computers, pp. 6-15, September 1993.

[13] M.A. Richards, "The Rapid Prototyping of Application Specific Signal Processors (RASSP) Program: Overview and Status", Proc. of Int'l Wshp on Rapid System Prototyping (RSP), Grenoble, France, June 1994.

[14] G.E. Fisher, "Rapid System Prototyping in an Open System Environment", Proc. of Int'l Wshp on Rapid System Prototyping (RSP), Grenoble, France, June 1994.

[15] C. Lavarenne, O. Seghrouchni, Y. Sorel, M. Sorine, "The SynDex software environment for real-time distributed systems design and implementation", Proc. European Control Conference, July 1991.

[16] P. Paulin, C. Liem, T. May, S. Sutarwala, "DSP Design Tool Requirements for Embedded Systems: A Telecommunications Industrial Perspective", to appear in Journal of VLSI Signal Processing (special issue on synthesis for real-time DSP), Kluwer Academic Publishers, 1994.

[17] V. Mooney, T. Sakamoto and G. De Micheli, "Run-time Scheduler Synthesis for Hardware-Software Systems and Application to Robot Control Design", In IEEE, Editor, Proceedings of the CHDL '97, pages 95-99, March 1997.

[18] J.-M. Daveau, G. Fernandes Marchioro, A.A. Jerraya, "VHDL Generation from SDL Specification", in Carlos D. Kloos and Eduard Cerny, editors, Proceedings of CHDL, pages 182-201-, IFIP, Chapman-Hall, April 1997.

[19] C. Delgado Kloos, Marin Lopez and all, "From LOTOS to VHDL", Current issues in Electronic Modelling, 3, September 1995.

[20] D. Hermann, J. Henkel, R. Ernst, "An Approach to the Adaptation of Estimated Cost Parameters in the COSYMA System, in Proc. Third Int'l Wshp on Hardware/Software Co-design Codes/CASHE, pages 100-107, Grenoble, IEEE CS Press, September 1994.

[21] J. Henkel, R. Ernst, "A Path-based Technique for Estimating Hardware Runtime in HW/SW-Cosynthesis", in 8th Intl Symposium on System Synthesis (ISSS), pages 116-121, Cannes, France, September 13-15 1995.

[22] D.C. Ku, G. De Micheli, "HardwareC – a Language for Hardware Design", Technical Report, N° CSL-TR-88-362, Computer Systems Lab, Stanford University, August 1998.

[23] W. Wolf, "Hardware-Software Co-design of Embedded Systems", Proceedings of IEEE, 27(1); 42-47, January 1994.

[24] L. Lavagno, A. Sangiovanni-Vincentelli, H. Hsieh, "Embedded System Co-design: Synthesis and Verification, pages 213-242, Kluwer Academic Publishers, Boston, MA, 1996.

[25] G. Berry, L. Cosserat, "The ESTEREL Synchronous Programming Language and its Mathematical Semantics", Language for Synthesis, Ecole Nationale Supérieure des Mines de Paris, 1984.

[26] G. Berry, "Hardware Implementation of Pure ESTEREL", in Proceedings of the ACM Workshop on Formal Methods in VLSI Design, January 1991.

[27] D. Gajski, F. Vahid, S. Narayan, J. Gong, "Specification and Design of Embedded Systems", Prentice Hall, 1994.

[28] D. Gajski, F. Vahid, S. Narayan, "System-Design Methodology: Executable-Specification Refinement", in Proc. European Design and Test Conference (EDAC-ETC-EuroASIC), pages 458-463, Paris, France, IEEE CS Press, February 1994.

[29] S. Narayan, F. Vahid, D. Gajski, "System Specification and Synthesis with the SpecCharts Language", in Proc. Int'l Conf. on Computer-Aided Design (ICCAD), pages 226-269, IEEE CS Press, November 1991.

[30] F. Vahid, S. Narayan, "SpecCharts: A Language for System-Level Synthesis", in Proceedings of CHDL, pages 145-154, April 1991.

[31] F. Vahid, S. Narayan, D. Gajski, "SpecCharts: A VHDL Front-End for Embedded Systems", IEEE Trans. on CAD of Integrated Circuits and Systems, 14(6): 694-706, 1995.

[32] ISO, IS 8807. "LOTOS: A Formal Description Technique based on the Temporal Ordering of Observational Behavior, February 1989.

[33] K. Van Rompaey, D. Verkest, I. Bolsens, H. De Man, "Coware – A Design Environment for Heterogeneous Hardware/Software Systems", in Proceedings of the European Design Automation Conference, Geneve, September 1996.

[34] R. Klein, "Miami: A Hardware Software Co-simulation Environment", in Proceedings of RSP'96, pages 173-177, IEEE CS Press, 1996.

[35] I. Bolsens, B. Lin, K. Van Rompaey, S. Vercauteren, D. Verkest, "Co-design of DSP Systems", in NATO ASI Hardware/Software Co-design, Tremezzo, Italy, June 1995.

[36] N.L. Rethman, P.A. Wilsey, "Rapid: A Tool for Hardware/Software Tradeoff Analysis", in Proc. IFIP Conf. Hardware Description Languages (CHDL), Elsevier Science, April 1993.

[37] M. Romdhani, R.P. Hautbois, A. Jeffroy, P. De Chazelles, A.A. Jerraya, "Evaluation and Com-
      position of Specification Languages, an Industrial Point of View", in Proc. IFIP Conf. Hardware
      Description Languages (CHDL), pages 519-523, September 1995.

[38] P. Paulin, J. Frehel, E. Berrebi, C. Liem, J.-C. Herluison, M. Harrand, "High-level Synthesis and
      Co-design Methods: An Application to a Video phone Codec", Invited Paper EuroDAC/VHDL,
      Bringhton, September 1995.

[39] D. Lanneer, J. VanPraet, W. Geurts, G. Goossens, "Chess: Retargetable code generation for
      embedded DSP processors", in Code Generation for Embedded Processors, ed. by P. Marwedel,
      G. Goossens, Kluwer Academic Publishers, 1995.

[40] P. Marwedel, G. Goessens, "Code Generation for Embedded Processors (DSP)", Kluwer Academic
      Publishers, 1995.

[41] A. Alomary, T. Nakata, Y. Honma, "An ASIP Instruction Set Optimization Algorithm with
      Functional Module Sharing Constraint", In ICCAD'93, pp. 526-532, 1993.

[42] R. Ernst, J. Henkel, Th. Benner, W. Ye, U. Holtmann, D. Herrmann, M. Trawny, "The COSYMA
      Environment for Hardware/Software Cosynthesis", Journal of Microprocessors and Microsystems,
      Butterworth-Heinemann, 1995.

[43] J. Madsen, J. Brage, "Co-design Analysis of a Computer Graphics Application", Journal: Design
      Automation of Embedded Systems, vol.1, no.1-2, January 1996.

[44] K. Buchenrieder, "A Prototyping Environment for Control-Oriented HW/SW Systems using State-
      Charts, Activity-Charts and FPGA's", Proc. Euro-DAX with Euro-VHDL, Grenoble, France, IEEE
      CS Press, pp. 60-65, September 1994.

[45] P. Camurati, F. Corno, P. Prinetto, C. Bayol, B. Soulas, "System Level Modeling and Verification:
      A Comprehensive Design Methodology", Proc. European Design & Test Conference, Paris, France,
      IEEE CS Press, February 1994.

[46] D. Gajski, F. Vahid, S. Narayan, "A System Design Methodology: Executable-Specification Refine-
      ment", Proc. European Design & Test Conference (EDAC-ETC-EuroASIC), Paris, France, IEEE
      CS Press, pp. 458-463, February 1994.

[47] M. Chiodo, D. Engels, P. Giusto, H. Hsieh, A. Jurecska, L. Lavagno, K. Suzuki, A. Sangiovanni-
      Vincentelli, "A case Study in Computer Aided Co-design of Embedded Controllers", Design
      Automation for Embedded Systems, Vol.1, No.1-2, pp. 51-67, Jan. 1996.

[48] W. Wolf, "Hardware-Software Co-Design of Embedded Systems", Proceedings of the IEEE, vol
      82. no 7, pp. 967-989, July 1994.

[49] J. Gong, D. Gajski, S. Narayan, "Software Estimation from Executable Specifications", Proc.
      European Design & Automation Conference (EuroDAC), IEEE CS Press, Grenoble, France,
      September 1994.

[50] D. Gajski, F. Vahid, "Specification and Design of Embedded Hardware-Software Systems", IEEE
      Design & Test of Computers, pp. 53-67, Spring 1995.

[51] M. Srivastava, R. Brodersen, "SIERA: A unified framework for rapid-prototyping of system-level
      hardware and software", IEEE Transactions on Computer-Aided Design of Integrated Circuits and
      Systems, pp. 676-693, June 1995.

[52] P. Paulin, C. Liem, T. May, S. Sutarwala, "DSP Design Tool Requirements for Embedded Systems:
      A Telecommunication Industrial Perspective", in Journal of VLSI Signal Processing (special issue
      on synthesis for real-time DSP), Kluwer Academic Publishers, 1994.

[53] A. Kalavade, E.A. Lee, "The extended Partitioning Problem: Hardware/Software Mapping,
      Scheduling, and Implementation-bin Selection", Proceedings of Sixth International Workshop on
      Rapid Systems Prototyping, North Carolina, pp. 12-18, June, 1995.

[54] C. Valderrama, A. Changuel, P.V. Raghavan, M. Abid, T. Ben Ismail, A. Jerraya, "A unified
      model for co-simulation and co-synthesis of Mixed Hardware/Software systems", The European
      Design and Test Conference ED & TC95 Paris, France, 6-9 March 1995.

[55] [SDL] ITU-T, 2.100 functional specification and description language, recommendation
      2.100-2.104, March 1993.

CHAPTER 8

TEST AND DESIGN-FOR-TEST: FROM CIRCUITS TO INTEGRATED SYSTEMS

MARCELO LUBASZEWSKI

Electrical Engineering Department, Universidade Federal do Rio Grande do Sul, UFRGS,
Av. Osvaldo Aranha esquina Sarmento Leite, 103, 90035-190 Porto Alegre RS, Brazil,
Phone: +55 51 3316-3516 Fax: +55 51 3316-3293. E-mail: luba@ece.ufrgs.br

Abstract: As demanding market segments require ever more complex, faster and denser circuits, high quality tests become essential to meet design specifications in terms of reliability, time-to-market, costs, etc. In order to achieve acceptable fault coverage for highly integrated systems, it is reasonable to expect that no other solution than design-for-test will be applicable in the near future. Therefore, testing tends to be dominated by embedded mechanisms to allow for accessibility to internal test points, to achieve on-chip test generation, on-chip test response evaluation, or even to make it possible the detection of errors concurrently to the circuit application. Within this context, an overview of existing test methods is given in this chapter, focusing on design-for-testability, built-in self-test and self-checking techniques suitable for digital and analog integrated circuits. Moreover, the application of these design-for-test techniques to integrated systems, implemented as multi-chip modules, microsystems or core-based integrated circuits, is also discussed

Keywords: Digital and analog test, fault modelling, fault simulation, test generation, design-for-test, design-for-testability, built-in self-test, self-checking circuits

1.   INTRODUCTION

Today's telecommunications, consumer electronics and other demanding market segments, require that more complex, faster and denser circuits are designed in shorter times and at lower costs. Obviously, the ultimate goal is to maximize profits.

However, the development of reliable products cannot be achieved without high quality methods and efficient mechanisms for testing circuits and integrated systems. To ensure the quality required for product competitiveness, one can no more rely on conventional functional tests: a move is needed towards methods that search for manufacturing defects and faults occurring during a system's lifetime. Moreover,

R. Reis et al. (eds.), Design of Systems on a Chip, 159–189.
© 2006 *Springer.*

to achieve acceptable fault coverage has become a very hard and costly task for external testing: test mechanisms need to be built into integrated circuits early in the design process. Ideally, these mechanisms should be reused to test for internal defects and environmental conditions that may affect the operation of the systems into which the circuits will be embedded. This would provide for amplified payback. To give an estimate about the price to pay for faults escaping the testing process, fault detection costs can increase by a factor of ten, when moving from the circuit to the board level, then from the board to the system level, and lastly, from the final test of the system to its application in the field.

Therefore, design-for-test seems to be the only reasonable answer to the testing challenges posed by state-of-the-art integrated systems. Mechanisms that allow for accessibility to internal test points, that achieve on-chip test generation, on-chip test response evaluation, and that make it possible the detection of errors concurrently to the application, are examples of structures that may be embedded into circuits to ensure system testability. They obviously incur penalties in terms of silicon overhead and performance degradation. These penalties must definitely account for finding the best trade-off between quality and cost. However, the industrial partic-ipation to recent test standardization initiatives confirms that, in many commercial applications, design-for-test can prove economical.

Within this context, the aim of this chapter is to give a glimpse into the area of design-for-test of integrated circuits and systems. Digital and analog circuits are considered. First of all, the test methods of interest to existing design-for-test techniques are revisited. Fault modelling, fault simulation and test generation are thus addressed. Then, design-for-testability, built-in self-test and self-checking tech-niques are discussed and illustrated in the realm of integrated circuits and printed circuit boards. Finally, the extension of these techniques to integrated systems is dis-cussed, considering multi-chip module implementations, microsystems embedding micro-electro-mechanical structures, and core-based integrated circuits.

2.   TEST METHODS

2.1   General Background

From the very first design, any circuit undergoes *prototype debugging*, *production* and periodic *maintenance tests* to simply identify and isolate, or even replace faulty parts. Those are called *off-line tests*, since they are independent of the circuit application and need, in the field, that the application is stopped before the related testing procedures can be run.

In high safety systems, such as automotive, avionics, high speed train and nuclear plants, poor functioning cannot be tolerated and detecting faults concurrently to the application becomes also essential. The *on-line detection* capability, used for checking the validity of undertaken operations, can be ensured by software and/or special mechanisms, such as watchdog timers, data encoding and self-checking hardware.

In general, tests must check whether, according to the specifications, the circuit has a correct functional behaviour (*functional testing*), or whether the physical implementation of the circuit matches its schematics (*structural testing*). The former, because it is run at-speed, can catch performance problems, such as excessive signal delays. The latter, opposed to the former, can provide a quantitative measure of test effectiveness (*fault coverage*), due to its fault-based nature. If, for any reason, structural tests cannot be run at-speed, they must be followed by functional tests, although this leads to a longer test time because of redundant testing.

On one hand, test methods have already reached an important level of maturity in the domain of digital systems: *digital testing* has been dominated by structured fault-based techniques and by successfully developed and automated standardized test configurations. On the other hand, practical analog solutions are still lagging behind their digital counterparts: *analog testing* has traditionally been achieved by functional test techniques, based on the measurement of circuit specification parameters such as gain, bandwidth, distortion, impedance, noise, etc. Considering the existing digital and analog testing techniques, it is not always easy to integrate them into a single *mixed-signal test* solution. Then, either ad hoc solutions for mixed circuits, such as CODECs and data converters, are developed, or analog and digital sections are partitioned in test mode, so that the inputs and outputs of each section can be accessed independently.

The application of input stimuli followed by the observation of circuit output voltages has been a widely used test technique. However, it has been shown that such a *voltage testing* cannot detect many physical defects that lead to unusual circuit consumptions. This is the reason why the practice of *current testing*, based on the measurement of the current consumption between the power supplies, has been increasing in importance in the last years. Current testing has been mostly faced as a complementary technique to the voltage testing approach.

2.2   Defects and Fault Models

Efficient tests can only be produced if realistic fault models, based on physical failure mechanisms and actual layouts, are considered.

Many defects may be inherent to the silicon substrate on which the integrated structures will be fabricated. Those may result from impurities found in the material used to produce the wafers, for example. Others may be due to problems occurring during the various *manufacturing* steps: the resistivity of contacts, for instance, will depend on the doping quality; the presence of dust particles in the clean room or in the materials may lead to the occurrence of spot defects; the misalignment of masks may result in deviations of transistor sizes, etc. All these defects lead, in general, to faults that simultaneously affect several devices, i.e. *multiple faults*.

Other defects occur during a circuit's *lifetime*. They are usually due to failure mechanisms associated to transport and electromechanical phenomena, thermal weakness, etc. In general, those defects produce *single faults*.

Permanent faults, like interconnect opens and shorts, floating gates, etc, can be produced by defects resulting from manufacturing and circuit usage. *Transient faults*, on the contrary, will appear due to intermittent phenomena, such as electro-magnetic interference or space radiations.

In the last decades, many models have appeared that tried to properly translate into faults actual physical defects (Wadsack, 1978; Galay, 1980; Courtois, 1981; Rajsuman, 1992). In the digital domain, gate inputs and outputs *stuck-at* a logical 0 or 1, always conducting (*stuck-on*) or never conducting (*stuck-open*) transistors, and slow-to-rise and slow-to-fall gates (*gate delays*) are examples of device fault models. Shorts (resistive or not) between wires (*bridging*), *stuck-open* wires, and the cumulative delay of gates and interconnects in the critical path (*path delay*) are examples of interconnect fault models. Specific fault models also exist for memories that include, for example, couplings of memory cells and transition faults. Obviously, highly complex microprocessors cannot be analysed under the lights of these fault models. Hence, those circuits rely on fault models that are based on the functionality of their instruction sets (*functional* fault model). In general, the existing fault models are complementary, in the sense that none can fully represent all possible physical defects.

Although defects are absolutely the same for digital and analog circuits, fault modelling is a much harder task in the analog case. This is mainly due to a larger number of possible misbehaviours resulting from defects that may affect a circuit dealing with analog signals. Three categories of faults have been guiding most works on analog testing: *hard faults* (Milor, 1989), *soft* and *large deviations* (Slamani, 1995). Hard faults are serious changes in component values or in circuit topology, usually resulting from device or wire opens and short circuits. Figure 1 shows an example of hard faults at the transistor level. Soft faults are small deviations around nominal values that cause circuit malfunctions. They may cause changes in the cut-off frequency of filters, changes in the output gain of amplifiers, etc. Large

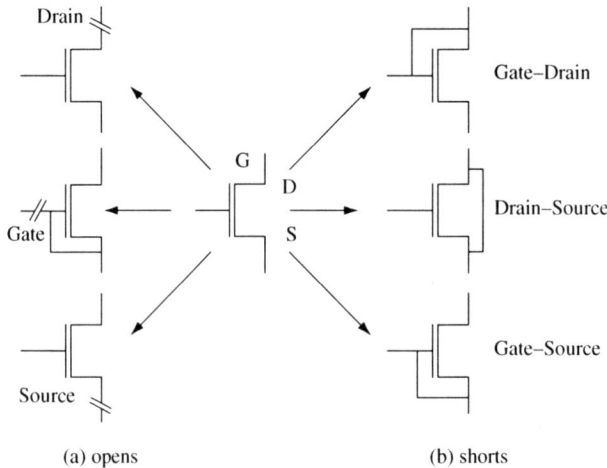

Figure 1. Fault model for hard faults in a MOS transistor

deviations are also deviations in the nominal value of components, but of a greater magnitude. They still cause circuit malfunctions, but their effects are quite harder than those observed for soft faults. Some few works also consider interaction faults in analog and mixed-signal circuits (Caunegre, 1996; Cota, 1997). These faults are shorts between nodes of the circuit that are not terminals of the same digital or analog component.

Finally, from the knowledge about defect sizes and defect occurrences in a process, layouts can be analysed and the probabilities of occurrence of opens, shorts, etc, in different layout portions and layers, can be derived by *inductive fault analysis* (Meixner, 1991). The IFA technique has the advantage of considerably reducing the list of faults, by taking into account only those with reasonable chances of occurrence (*realistic faults*).

2.3   Fault Simulation

Fault simulation consists, basically, of simulating the circuit in the presence of the faults of the model, and of comparing the individual results with the fault-free simulations. The goal is to check whether or not these faults are detected by the applied input stimuli. The steps involved in the fault simulation process are:
– fault-free simulation;
– reduction of the fault list (*fault collapsing*), by deleting faults that present equivalent behaviour (*fault equivalence*) and faults that dominate other faults (*fault dominance*);
– insertion of the fault model into the fault-free circuit description (*fault injection*);
– simulation of the faulty circuit;
– comparison of the faulty and fault-free simulation results and, in case of mismatch, deletion of the fault from the initial fault list (*fault dropping*).
A typical fault simulation environment is given in Figure 2.
Fault simulation is widely used for:
– test evaluation, by checking the fault coverage, i.e. the percentage of faults detected by a set of input stimuli;
– fault dropping in test generation, by verifying which faults of the model a computed test pattern detects; and
– diagnosis, by making it possible the construction of *fault dictionaries* that identify which faults are detected by every test vector.
Many digital fault simulation techniques exist that are based on a single stuck-at fault model: In the *parallel fault simulation* approach, the fault-free circuit and a limited number of faulty circuits are simulated in parallel (Seshu, 1965). Packed data structures are used that store the good and all bad circuit signal values. In the *deductive fault simulation* (Armstrong, 1972), only the fault-free circuit is simulated, while the behaviour of a fixed number of faulty circuits is deduced. A fault list, associated to every circuit node, keeps track of all faults at the current simulation time, that cause that node to behave differently in the fault-free and in the faulty circuit. Finally, in the *concurrent fault simulation* approach (Ulrich, 1974), the fault-free circuit is fully simulated, and only the elements that are different from those

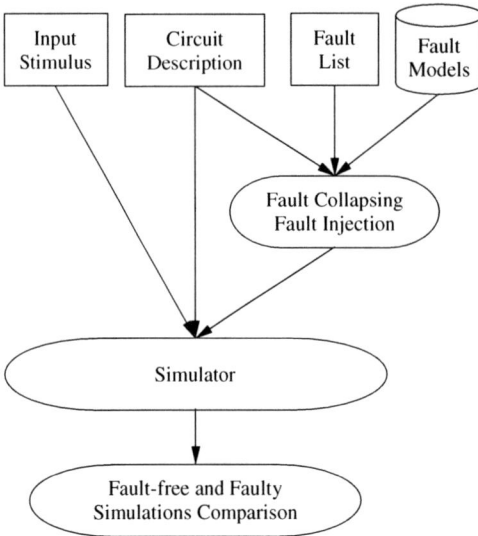

Figure 2. A generic fault simulation procedure

of the fault-free circuit are simulated in the faulty circuits. More recent research has mainly focused on special-purpose hardware accelerators and on algorithms for hierarchical fault simulation (Abramovici, 1990).

Contrarily to the digital case, where some degree of parallelism is possible in fault simulation, in analog circuits faults are injected and simulated sequentially, making analog fault simulation a very time consuming task. For analog circuits, fault simulation is traditionally performed at the transistor-level using circuit simulators. For the fault simulation of continuous-time analog circuits, (Sebeke, 1995) proposes a computer-aided testing tool based on a transistor-level hard fault model. This fault model is made up of local shorts, local opens, global shorts and split nodes. The tool injects into the fault-free circuit realistic faults obtained from layouts by an IFA-based fault extractor. For switched-capacitor analog circuits (Mir, 1997) presents a switch-level fault simulator that models shorts, opens and deviations in capacitors, stuck-on, stuck-open and shorts between analog terminals of switches. An automatic tool is introduced that performs time- and frequency-domain switch-level fault simulations, keeping simulation times orders of magnitude lower than for transistor-level simulations. A behavioural-level fault simulation approach is proposed in (Nagi, 1993a). It has practical use only for continuous-time linear analog circuits. First of all, the circuit under test, originally expressed as a system of linear state variables, suffers a bilinear transformation from the s-domain equations to the z-domain. Next, the equations are solved to give a discretized solution. In this approach, soft faults in passive components are directly mapped onto the state equations, while hard faults require, in general, that the transfer function is recomputed for the affected blocks. Finally, operational amplifier faults are modelled in the s-domain, before mapping them to the z-domain.

2.4   Test Generation

Following the choice of a suitable fault model and fault simulator, test generation is the natural step to define an efficient test procedure to apply to the circuit under test.

   The problem of generating tests consists, basically, of finding a set of input test stimuli and a set of output measures, which guarantee maximum fault coverage. If the fault detection goal is extended to include fault diagnosis, the computed test stimuli must, additionally, be capable of distinguishing between non-equivalent faults. A typical test generation environment is given in Figure 3.

   In the digital domain, test pattern generation can be achieved by means of exhaustive, pseudorandom or deterministic approaches. *Exhaustive* testing stands for the generation and application of all possible patterns to a circuit inputs. Although generating exhaustive tests is a trivial task and ensures the highest fault coverage, this approach is only practical for very small circuits. Special test generators can produce *pseudorandom* patterns considering different seeds and different length requirements for the test sequences. For random logic, this method provides, in general, very high fault coverage within reasonable test application times. Nevertheless, in many circuits that need just some few vectors to achieve high fault coverage, the use of the pseudorandom approach would result in too long test sequences. This is the

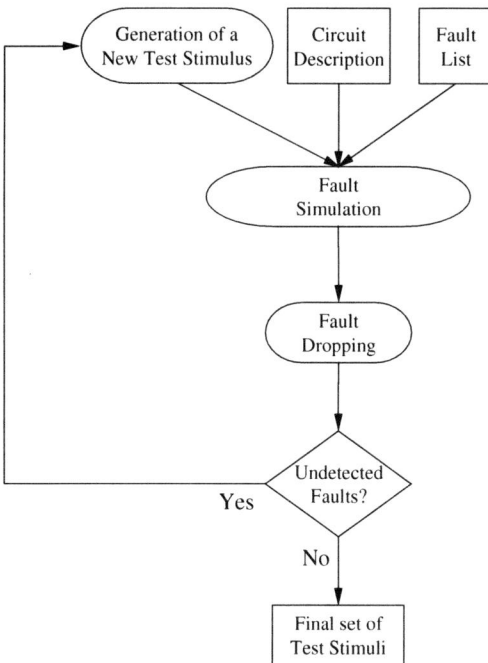

Figure 3. A generic test generation procedure

case of circuits based on regular structures. To solve this problem, much shorter test sequences can be precomputed by means of *deterministic* test pattern generators.

Two types of methods address test pattern generation in a deterministic way: algebraic and topological methods. In *algebraic methods*, test vectors are computed from the Boolean expressions that describe the circuit under test. In *topological methods*, test patterns are generated based on the circuit structure. The far most successful test generation approach is undoubtedly the *D-algorithm* (Roth, 1967), based on a topological method. In addition to 0's and 1's, the D-algorithm uses the values D and /D (complement of D) to distinguish fault-free and faulty circuit behaviours. D (/D) denotes 1(0) as the correct behaviour and 0(1) as the faulty node behaviour. Essentially, the D-algorithm starts by propagating a single stuck-at fault (denoted by D or /D) to a circuit primary output. Next, it tries to justify the primary inputs, such that the D's and /D's computed in the previous step are actually obtained within the circuit. The D-algorithm was firstly developed to deal with combinational circuits, and later on extended to generate patterns for sequential circuits. Some examples of well-established test pattern generators are PODEM (Goel, 1981) and FAN (Fujiwara, 1983), both based on improvements of the D-algorithm. More recent contributions have specially addressed hierarchical test pattern generation, test generation from high-level circuit descriptions (Murray, 1996), test vector generation for sequential circuits (Marchok, 1995), and test generation considering other fault models, such as delay faults (Brackel, 1995), and other testing methods, such as current testing (Nigh, 1989).

Over the last years, some test generation procedures for analog circuits have been proposed in the literature. The technique reported in (Tsai, 1991), one of the earliest contributions to test generation of linear circuits, formulates the analog test generation task as a quadratic programming problem, and it derives pulsed waveforms as input test stimuli. DC test generation is dealt with in (Devarayanadurg, 1994) as a minmax optimization problem that considers process variations for the detection of hard faults in analog macros. This minmax formulation of the static test problem is extended to the dynamic case (AC) in (Devarayanadurg, 1995). The automatic generation of AC tests has also been addressed in other works (Nagi 1993b; Slamani, 1995; Mir, 1996a; Cota, 1997). (Nagi, 1993b) uses a heuristic based on sensitivity calculations to choose the circuit frequencies to consider. After each choice, fault simulation is performed as the means to drop from the fault list all detected faults. From a multifrequency analysis, the approach in (Slamani, 1995) selects the test frequencies that maximize the sensitivity of the output parameters measured for each individual faulty component. Mir (1996a) also proposes a multifrequency test generation procedure, but computes a minimal set of test measures and a minimal set of test frequencies which guarantee maximum fault coverage and maximal diagnosis. Finally (Cota, 1997) enlarges the set of faults including interaction shorts, and merges the sensitivity analysis (Slamani, 1995) and the search of minimal sets (Mir, 1996a), with test generation based on fault simulation (Nagi, 1993b). Additionally, it applies the new automatic test generation procedure to linear and non-linear, and to analog and mixed-signal circuits.

3.   DESIGN-FOR-TEST

Even though a test generation tool is available for testing, hard-to-detect faults can prevent that a good trade-off between fault coverage and testing time is achieved. In these cases, the redesign of parts of the circuit can represent a possible solution to improve the accessibility to hard-to-test elements (*design-for-testability*).

Considering the increasing complexity of integrated circuits, another design-for-test possibility is to build self-test capabilities into circuits (*built-in self-test*). In general, the use of on-chip structures for test generation and test evaluation allows for significant savings in test equipment, reducing the final chip cost.

In case the application requires that faults are detected on-line, the circuit can be made *self-checking* by encoding the outputs of functional blocks and verifying them through embedded checkers. Unlike built-in self-test approaches, concurrent checking is performed using functional signals, rather than signals specifically generated for testing the circuit.

In the following sections, these three design-for-test approaches, i.e. design-for-testability, built-in self-test and self-checking technique, are further discussed and illustrated. The discussion ends up by the proposal of an unified approach for off-line and on-line testing of integrated circuits.

3.1   Design-for-Testability

Design-for-testability approaches aim at improving the capability of observing, at the circuit outputs, the behaviour of internal nodes (*observability*), and at improving the capability of getting test signals from the circuit inputs to internal nodes (*controllability*).

Ad-hoc techniques, that are in general based on partitioning, on the use of multiplexers to give access to hard-to-test nodes, on disabling feedback paths, etc, can be used to enhance the testability of circuits. Nevertheless, structured approaches are far more suitable to face testing problems in highly complex integrated circuits.

In the digital domain, the most successful structured approach is undoubtedly the *scan path* technique (Eichelberger, 1978). This technique makes use of scan, in order to improve the observability and the controllability of nodes internal to sequential circuits. In test mode, a set of circuit flip-flops is connected into a shift register configuration, so that scan-in of test vectors and scan-out of test responses are made possible. Similarly, the testability of internal nodes of printed circuit boards can be improved by extending the internal scan path to the interface of integrated circuits. This technique is referred to as *boundary scan* (LeBlanc, 1984). Figure 4 shows an ideal boundary scan (BS) board and a possible implementation for the scan cells required at the circuit boundaries. Note that the boundary scan cells are connected to each other in series, in order to build a circuit and, subsequently, a board scan path. By properly playing with the modes of multiplexers of boundary scan cells, three different types of tests can be applied to the board (Maunder, 1990): an external test, based on the control and observation of board interconnects; an internal test, consisting of the control of the inputs and the observation of the outputs of integrated circuits; and a sample test, based on the observation of both integrated circuit inputs and outputs.

The boundary scan technique is the basis of a very successful test standard (IEEE, 1990), implemented in many products available in the market (Maunder, 1994). The architecture of the IEEE Std. 1149.1 is made up of an instruction register, test data registers (the boundary scan register, for example), a test access port (TAP) and its controller. The TAP is composed of 4 mandatory pins: TDI (test data input), TDO (test data output), TMS (test mode select) and TCK (test clock). The TAP controller consists of 16 states providing for register selection, capture, shift (with or without pause) and update, for test logic reset and self-test run.

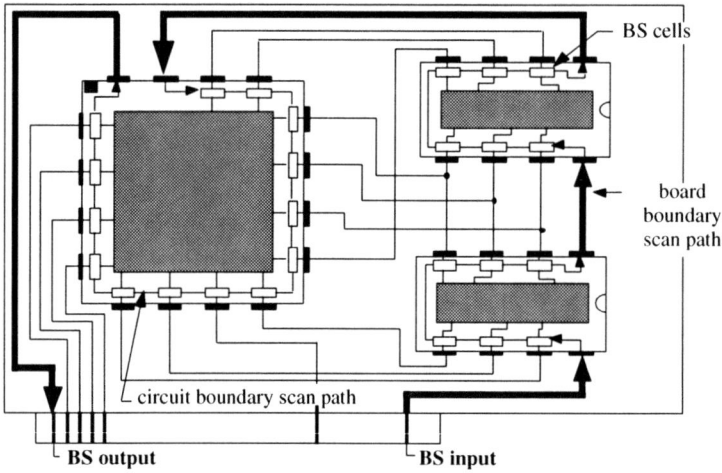

(a) boundary scan path

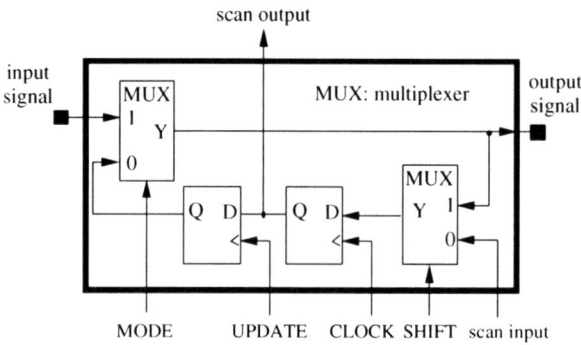

(b) boundary scan cell

Figure 4. Boundary scan testing

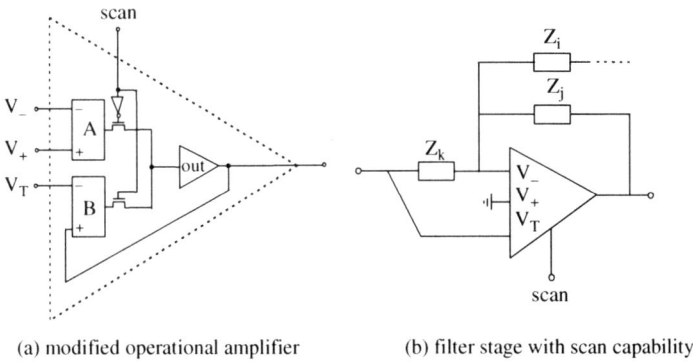

(a) modified operational amplifier   (b) filter stage with scan capability

Figure 5. Analog scan based on operational amplifier with duplicated input stage

In the analog case, the first attempt to apply the idea of the scan path to test filters was made by (Soma, 1990). The basis of this design-for-testability methodology consists of dynamically broadening the bandwidth of each stage of a filter, in order to improve the controllability and observability of circuit internal nodes. This band width expansion is performed by disconnecting the capacitors of the filter stages by using MOS switches. The main drawback of this approach is that the additional switches impact the filter performance. Although in the extension of this technique to switched-capacitor implementations (Soma, 1994) no extra switches are needed in the main signal path, additional control signals and associated circuitry and routing are required. More recently, operational amplifiers with duplicated input stages have been used, in order to reduce the impact on performance of the additional scan circuitry (Bratt, 1995). This technique is illustrated in Figure 5: in scan mode, a filter stage can be reprogrammed to work as a voltage follower, and propagate to the next stage the output of the previous stage.

At the board level, the main problem to face in testing mixed-signal circuits is the detection and diagnosis of interconnect faults. While shorts and opens in digital wiring can be easily checked by means of boundary scan, analog interconnects (made up of discrete components in addition to wires) require specific mechanisms to measure impedance values. The principle of the most usual impedance mea-surement technique is shown in Figure 6 (Osseiran, 1995). Zx is the impedance to measure. $Z1$ and $Z2$ are two other impedances connected to Zx. Zs is the output impedance of the test stimuli source (including the probe impedance), Zi is the input impedance of the measuring circuitry (impedance of the probe to the virtual ground of the operational amplifier) and Zg is the impedance of the probe connecting $Z1$ and $Z2$ to ground. If Zs, Zi and Zg are very low, Zx will be given by the formula in Figure 6.

Assuming that the terminals of Zx, $Z1$ and $Z2$ are connected to chips, electronic access to these points can be achieved by building into I/O interfaces the *analog boundary scan* cell given in Figure 7. While applying the measurement procedure

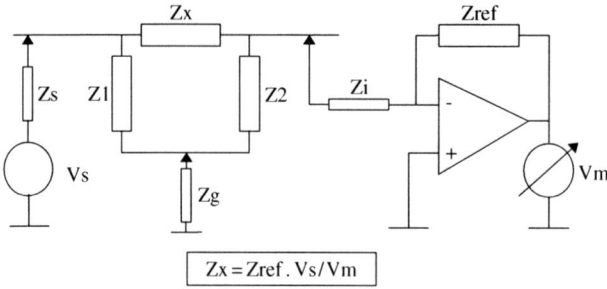

$$Zx = Zref . Vs / Vm$$

Figure 6. Analog in-circuit test

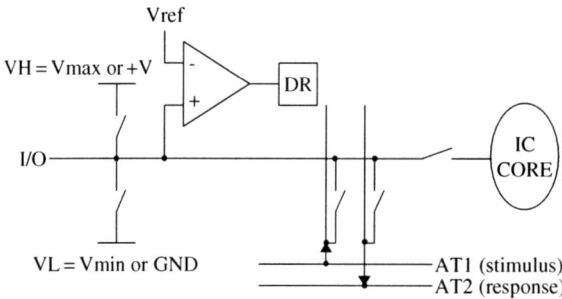

Figure 7. Analog boundary scan cell

described above, those cells to which Zx is connected will switch on bus AT1 for stimulus application, and switch on bus AT2 for response measurement. The cells connected to the ends of Z1 and Z2 that are opposed to Zx will be pulled down by switching on VL. All analog boundary scan cells will provide for isolation of integrated circuit cores. The measurement technique in Figure 6 and the analog cell in Figure 7 are part of the P1149.4 standard for a mixed-signal test bus (Osseiran, 1995).

This test standard shall extend the IEEE Std. 1149.1 to cope with the test of analog circuitry.

3.2   Built-in Self-test

With the advances on integrated circuits, faster and more complex test equipments are needed to meet ever more demanding test specifications. Testers with demanding requirements of speed, precision, memory and noise are in general very expensive. An attractive alternative is to move some or all of the tester functions onto the chip itself or onto the board on which the chips are mounted. The use of built-in self-test (BIST) for high volume production of integrated circuits is desirable to reduce the cost per chip during production-time testing.

An ideal BIST scheme should provide means of on-chip test stimulus generation, on-chip test response evaluation, on-chip test control and I/O isolation. The interest in a particular approach depends on its suitability to address the circuit faulty behaviours, and the cost and applicability of the technique.

All BIST methods have some associated cost in terms of area overhead and additional test pins. The additional BIST area required in the chip results in a decrease in yield. This penalty must be compensated by reducing test and maintenance costs. Moreover, by adding path delays, the BIST circuitry can degrade the circuit performance.

Ideally, a BIST structure would be applicable for any kind of functional circuit. The diversity in design architectures and functional and parametric specifications prevents reaching this aim. However, some structured approaches are applicable to wide classes of circuits. The interest in a BIST technique is also related to the ability to perform diagnosis in the field and the possibility of reusing hardware already available in the functional design.

3.2.1   Test generation and test compaction

Several BIST approaches have been proposed in the last years that are now common practice among digital designers.

For on-chip test generation, for example, deterministic patterns can be pre-stored in a ROM (Dandapani, 1984), or they can be generated by counters and decoders (Akers, 1989). In general, deterministic test generators are not practical yet due to the high hardware overheads obtained at the chip level. Pseudorandom pattern generators, on the contrary, require much lower overheads, in spite of much longer test application times. Most pseudorandom generators employed in industrial designs are based on *linear feedback shift registers* – LFSRs (Bardell, 1987). A general LFSR is composed of D type flip-flops connected as a shift register and of exclusive-OR gates implementing the linear feedback network. More recent works take advantage from the combination of deterministic and pseudorandom test generators (Dufaza, 1991).

In terms of on-chip test evaluation, compaction methods are often required, since usually it is not practical to pre-store all expected test responses in an embedded memory. Regardless the compaction method used, the inherent loss of information can lead to identical test results for a faulty and the fault-free circuit (*aliasing*). Existing test compaction methods are based on counting output transitions from 0 to 1 or from 1 to 0 (Hayes, 1976), on counting the number of 0's appearing in the output signal (Savir, 1980), on polynomial divisions implemented by LFSRs (Benowitz, 1975), or on digital integration suitable for accumulator-based architectures (Rajski, 1992). Most applications use the polynomial division method, also referred to as *signature analysis*. Basically, this method consists on dividing a polynomial that represents the set of circuit outputs by another polynomial that characterizes the response compaction LFSR. After the test application, the signature in the LFSR corresponds to the remainder of the polynomial division.

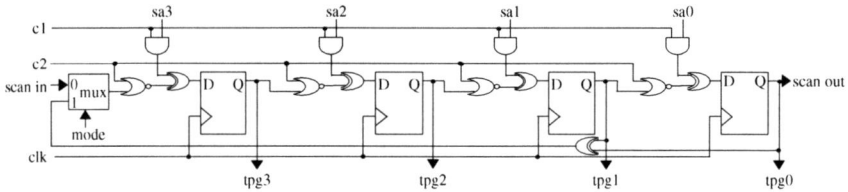

Figure 8. A 4-input BILBO

The merger of at-speed built-in test generation and signature analysis, with the scan path technique, culminated in the past with the proposal of a multifunctional BIST structure named BILBO: Built-In Logic Block Observer (Koenemann, 1979). A BILBO is basically a LFSR allowing for programmable operation. Four different functional modes can be selected in a digital BILBO (Figure 8). The structure can work as a latch (normal mode, $c1 = c2 = 1$), as a linear shift register (scan-based test mode, $c1 = c2 = mode = 0$), as a multiple-input signature register or a pseudorandom test pattern generator (BIST mode, $c1 = mode = 1$, $c2 = 0$), and it can be reset (initialisation mode, $c1 = 0$, $c2 = 1$). Currently, the BILBO structure is probably the most widespread BIST technique in use by digital designers.

In the realm of analog circuits, some few works have recently proposed on-chip structures for test generation and signature analysis.

In general, the stimulus generation for analog BIST depends on the type of test measurement to apply (Mir, 1995): DC static, AC dynamic or time domain measurements. DC faults are usually detected by a single set of steady state inputs; AC testing is typically performed using sinewave forms with variable frequency; finally, pulse signals, ramps or triangular waveforms are the input stimuli for time domain measurements. Relaxation and sinewave oscillators (Gregorian, 1986) are used for the generation of test signals. Dedicated sinewave oscillators have already been proposed for multifrequency testing (Khaled, 1995). To minimise the test effort, individual test signals can be combined to form a multitone test signal (Lu, 1994; Nagi, 1995). To save hardware, a method to reconvert a sigma-delta D/A converter into a precision analog sinewave oscillator has been proposed in (Toner, 1993). A different way of generating analog test stimuli consists on feeding pseudorandom digital test patterns into a D/A converter. In this method, called hybrid test stimulus generator (Ohletz, 1991), the digital patterns are generated as in a digital BILBO.

For analog circuits, the analysis of the output response is complicated by the fact that analog signals are inherently imprecise. The analysis of the output response can be done by matching the outputs of two identical circuits. This is possible if the function designed leads to replicated sub-functions or because the circuit is duplicated for concurrent checking (Lubaszewski, 1995). When identical outputs are not available, three main approaches can be considered for analysing the test response (Mir, 1995): In the first approach, the analog BIST includes analog checkers which verify the parameters associated with the analog behaviour (according

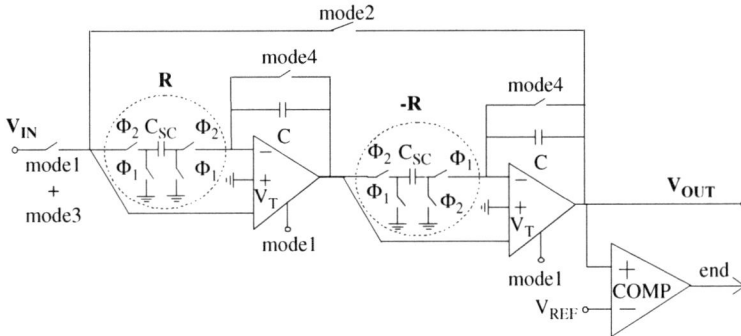

Figure 9. A switched-capacitor analog BILBO

to the specification) for known input test signals (Slamani, 1993). The second approach consists on the generation of a signature that describes the waveform of the output response. Recently, a compaction scheme that uses a digital integrator has been reported (Nagi, 1994). The third approach is based on the conversion of the analog test responses into digital vectors which, similarly to a digital BILBO, are fed into a signature analysis register to generate a signature (Ohletz, 1991).

The ability of scanning signals and generating/compacting analog AC-tests using the same hardware has recently led to the proposal of a novel multifunctional BIST structure. This structure, called analog built-in block observer (ABILBO), recreates the digital BILBO versatility in the analog domain (Lubaszewski, 1996). Basically, the ABILBO structure is made up of two analog integrators and one comparator. A switched-capacitor implementation is given in Figure 9. Since integrators have duplicated input stages as in Figure 5, the operational amplifiers can work as voltage followers and then perform analog scan (mode1). With the operational amplifiers in the normal mode, switches can be properly programmed, such that either a sinewave oscillator (mode2) or a double-integration signature analyser (mode3) results. The frequency of the quadrature oscillator obtained in mode2 depends linearly on the frequency of the switching clock ($\Phi 1, \Phi 2$). The signature resulting from the selection of mode3 in the ABILBO structure corresponds to the time for the output of the second integrator to reach a predefined reference voltage (V_{REF}). If a counter is used for computing digital signatures, counting must be enabled from the integration start up to the time when the comparator output goes high. In (Renovell, 1997), the ABILBO mode for signature analysis is extended to cope with transient tests. Finally, both integrators can be reset by shorting their integration capacitors (mode4).

3.2.2   Current testing

Many faults, such as stuck-on transistors and bridging faults, result in higher than normal currents flowing through the power supplies of the circuit under test (Maly, 1988).

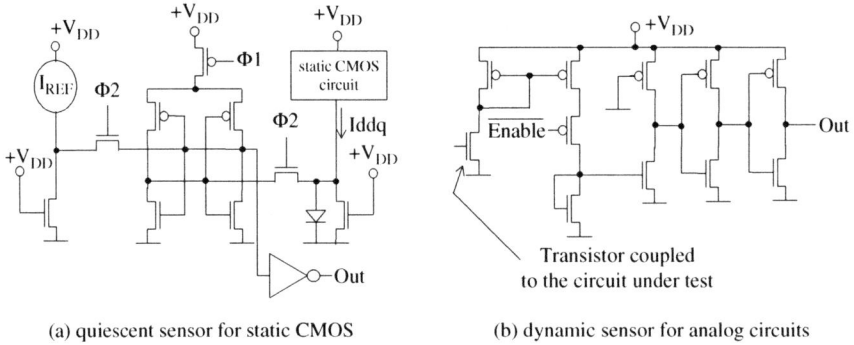

(a) quiescent sensor for static CMOS   (b) dynamic sensor for analog circuits

Figure 10. Built-in current sensors

In the case of digital CMOS circuits, for example, these faults create a path between V_{DD} and G_{ND} that should not exist in the fault-free circuit. Since the quiescent current becomes orders of magnitude higher than the expected leakage currents, these faults can be detected by using off-chip current sensors. This test method simplifies the test generation process, since the propagation of faults to the circuit primary outputs is no more required. In order to lower the evaluation time of the off-chip approach, intrinsically faster *built-in current sensors* can be used. Figure 10a shows an example of built-in current sensor. This sensor was proposed by (Lo, 1992) and consists, essentially, of a sense amplifier that compares the current of the circuit under test to a reference. The sensor is made very fast due to the use of two non-overlapping clock phases ($\Phi 2$ for sensing and $\Phi 1$ for evaluation), and due to a shunt diode used to draw off large transient current spikes and to limit the sampling voltage to 0.65 volts.

In the analog world, the same test method may apply to those circuits that present medium to low quiescent currents. For circuits with high quiescent currents, a possibility is to measure transients using specific *built-in dynamic current sensors*. The sensor proposed in (Argüelles, 1994) is shown in Figure 10b. It can be used to measure the dynamic current across the most sensitive branches of the circuit under test. To avoid performance degradation, this sensor is coupled to the circuit by means of an additional stage to existing current mirrors. As it can be seen from Figure 10b, in test mode (Enable = 1), the transient current is firstly copied, next converted to a voltage and amplified, and finally digitised. The sensor outputs a signature characterized by the number and width of pulses fitting a predefined time window.

Potentially, methods based on current measurements can lead to unified solutions for testing digital and analog parts of mixed-signal integrated circuits (Bracho, 1995).

3.2.3   Thermal testing

The advance of the technology makes today thermal issues ever more serious, due to the decreasing feature size of integrated circuits and the increasing density of packaging. This increasing density makes heat removal much harder, leading to a greater number of failures in the field.

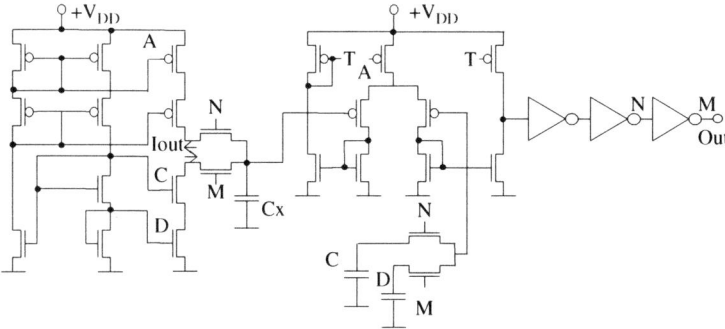

Figure 11. Current-based thermal sensor with frequency-output

To cope up with this problem, *thermal sensors* can be built into integrated circuits. They can thus be used for production and periodical off-line testing, as well as for on-line temperature monitoring. During the system use, if an embedded thermal sensor indicates overheating, the system itself can, for example, decrease the operating frequency or switch off functional parts. The goal is clearly to lower the temperature and avoid circuit failures. This technique is referred to as design for thermal testability (Szekély, 1995).

The types of temperature sensors that are of interest for thermal testing are thermoresistors, thermocouples, bipolar transistors and diodes, and MOS transistors (Gardner, 1994). An example of CMOS sensor with digital output is given in Figure 11 (Szekély, 1998). This sensor produces an analog current (I_{OUT}) proportional to the temperature. This current is next converted into digital information by means of a current-frequency converter. This sensor requires a very low silicon area and a low power consumption. Additionally, it presents a weak V_{DD} dependence and a good long-term stability.

3.3   Self-checking Circuits

In digital self-checking circuits, the concurrent error detection capability is achieved by means of functional circuits, which deliver encoded outputs, and checkers, which verify whether these outputs belong to error detecting codes. The most usual codes are the parity, the Berger and the double-rail code. The general structure of a self-checking circuit is shown in Figure 12.

Most often, self-checking circuits are aimed at reaching the *totally self-checking goal*: the first erroneous output of the functional circuit results in an error indication in the checker outputs. The basic properties required for achieving this goal are independent of the circuit implementation. They can be described at an abstract level in terms of the fault-free and the faulty functions of the circuit. These properties have been initially defined in (Carter, 1968) and further refined in (Anderson, 1971; Smith, 1978; Nicolaidis, 1988a). In general, the effectiveness of totally self-checking

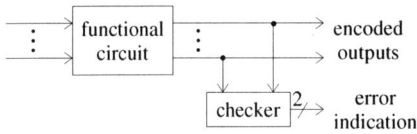

Figure 12. Self-checking circuit

circuits is based on the hypothesis that faults occur one at a time, and that, between any two faults, enough time elapses so that the functional circuit and the checker receive all the elements of the respective input code spaces.

In the literature, many works have addressed the design of *self-checking* circuits: *PLAs* (Nicolaidis, 1991), *RAMs* (Nicolaidis, 1994), *adders* and *ALUs* (Nicolaidis, 1993a), *microprocessor control parts* (Nicolaidis, 1990), etc. The surface overhead penalty reported in these works ranges from 10 to 84%.

Similarly to digital self-checking circuits, the aim of designing *analog self-checking circuits* is to meet the totally self-checking goal. This is possible since analog codes can also be defined, for example the differential and duplication codes (Kolarík, 1995). A tolerance is required for checking the validity of an analog functional circuit and this is taken into account within the analog code.

The nodes to be monitored by an analog checker are not necessarily those associated with the circuit outputs, due to commonly used feedback circuitry. In addition, the most important difference is that the input and output code spaces of an analog circuit have an infinite number of elements. Therefore, the hypothesis considered for digital circuits becomes unrealistic, since an infinite number of input signals might be applied within a finite lapse of time. In order to cope with this problem, the self-checking properties are redefined for the analog world in (Nicolaidis, 1993b).

In the last few years, the self-checking principle has been applied to on-line testing analog and mixed-signal circuits, including filters and A/D converters (Lubaszewski, 1995). The major techniques employed for concurrent error detection are: *partial replication* of modular architectures (Huertas, 1992), e.g. filters based on a cascade of biquads and pipelined A/D converters; *continuous checksums* in state variable filters (Chatterjee, 1991); *time replication* in current mode A/D converters (Krishnan, 1992); and *balance checking* of fully differential circuits (Mir, 1996b). Figure 13 illustrates the principle of balance checking applied to fully differential integrated filters. In a correctly balanced fully differential circuit, the operational amplifier inputs are at virtual ground. But, in general, transient faults, deviations in passive components and hard faults in operational amplifier transistors corrupt this balance. In (Mir, 1996b), an analog checker is proposed which is capable of signalling balance deviations, i.e. the occurrence of a common-mode signal at the inputs of fully differential operational amplifiers. This same technique was used for on-line testing A/D converters in (Lubaszewski, 1995) and in (Francesconi, 1996).

3.4   Unified Built-in Self-test

Faults originating from an integrated circuit's manufacture typically manifest as multiple faults. However, conventional self-checking architectures only cover single faults. Besides that, fault latency may lead to the accumulation of faults and can invalidate the self-checking properties. In addition, when the checkers generate an error indication in these circuits, no mechanism exists to recognize if the detected fault is a transient or a permanent one. But this information is important to allow for diagnosis and repair in the field.

A solution to these problems has been given in (Nicolaidis, 1988b). Nicolaidis proposes that built-in self-test capabilities similar to those used for production testing are embedded into self-checking circuits. These capabilities must be repeatedly activated, at periods of time no longer than the mean-time between failures. This technique, referred to as *unified built-in self-test (UBIST)*, unifies *on-line and off-line tests*, covering all tests necessary during a system's lifetime: manufacturing, field testing and concurrent error detection. Moreover, it simplifies the design of checkers and increases the fault coverage of self-checking circuits.

The general structure of the UBIST scheme is given in Figure 14. The off-line test pattern generation and signature analysis capabilities are achieved by the *UBILBO* modules. These modules are made up of a BILBO-like part and a coding part. The coding part complements the test stimuli generation by encoding test vectors and providing a code/noncode indication associated to test patterns. Two test phases are required in the UBIST scheme: during test 1 (test 2) phase, test vectors generated by odd (even) UBILBOs are applied to odd (even) functional circuits and to even (odd) checkers. The responses of odd (even) functional circuits are verified by odd (even) checkers and compacted by even (odd) UBILBOs to give a signature. At the end of test 1 (test 2), the signatures in even (odd) UBILBOs are shifted out for verification.

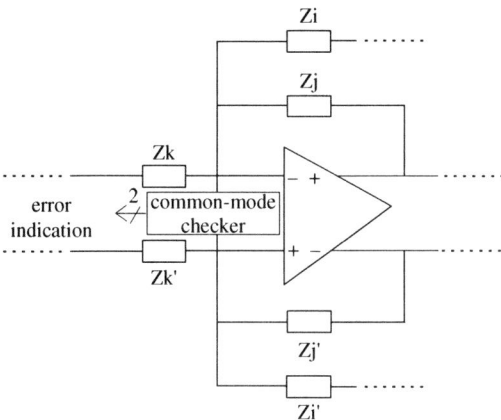

Figure 13. Generic stage of a self-checking fully differential filter

Similarly to integrated circuits, the unification of off-line and on-line tests
has also been proposed for printed circuits boards (Lubaszewski, 1992). This technique basically merges the UBIST scheme with boundary scan. More recently, this
merger has been further extended, and it has been applied to fault tolerant systems
based on duplicated self-checking modules with fail-safe interfaces (Lubaszewski,
1998a).

In the analog domain, the first attempt to couple built-in self-test and self-
checking capabilities was made by (Mir, 1996c). Mir proposes the design of a *test
master* compliant with the IEEE boundary scan standard, that efficiently shares
hardware between the *off-line and on-line tests of fully differential circuits*. This
test master relies on a programmable sinewave oscillator for test generation and
on common-mode analog checkers for test response evaluation. The frequencies to
apply to the circuit under test are computed by the test generation tool described
in (Mir, 1996a). For concurrent error detection, the checkers monitor the balance
of the inputs of fully differential operational amplifiers. To allow for off-line fault
detection and fault diagnosis, they additionally observe the balance of operational
amplifier outputs (Mir, 1996b).

In analog circuits built from a *cascade of similar functional modules*, the on-line
testing capability can be ensured by an additional checking module and a multiplexing system. The *programmable checking module* must be capable of mimicking
every functional module in the circuit. The *multiplexing system* must be such that
the outputs of every module can be compared against the outputs of the checking
module, when the latter receives the same inputs as the former. Assuming an analog
filter based on a cascade of biquads, the partial replication illustrated in Figure 15
can ensure that on-line tests test 1, test 2 and test 3 are applied, in a time-shared
manner, to individual biquads. The individual biquads can be designed such that
they can accommodate, in *off-line test* mode, the *ABILBO structure* of Figure 9.
Then, off-line tests can be applied in three different phases: In phase test 1, biquad 1
will be tested with biquad 3 working as an oscillator (ABILBO 3) and biquad 2
working as a signature analyser (ABILBO 2). In phase test 2, biquad 2 will be
tested with biquad 1 working as an oscillator (ABILBO 1) and biquad 3 working

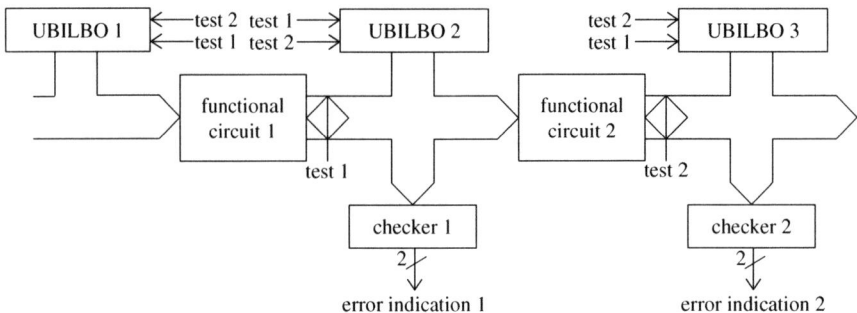

Figure 14. Unified built-in self-test

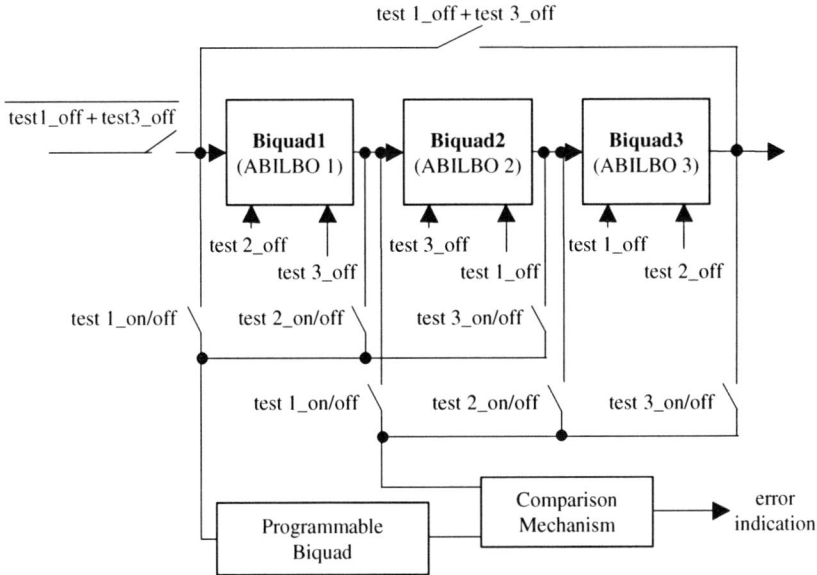

Figure 15. On-line/off-line test merger in a modular analog filter

as a signature analyser (ABILBO 3). In phase test 3, biquad 3 will be tested with biquad 2 working as an oscillator (ABILBO 2) and biquad 1 working as a signature analyser (ABILBO 1). A feedback path from the output to the filter input is required to apply the phases test 1 and test 3. In summary, the biquads, while working as test generators, test individual filter stages off-line, and check the ability of the programmable biquad to mimmick the filter stages on-line. While working as signature analysers, the biquads check that the test generators work properly, at the same time as they improve the fault diagnosis capability. This occurs because they make it possible to recognize if a fault affects the stage under test or the programmable biquad. As illustrated by this example, the unification of off-line and on-line tests in modular analog circuits is, in general, expected to result in low performance degradation and low overhead penalty.

4.   TESTING INTEGRATED SYSTEM

For the design of high performance and high density electronic systems, an alternative to the printed circuit board technology can be found in *multi-chip modules.* These single package modules consist of multiple dies interconnected in a high density substrate. Unlike large single chips, multi-chip modules can support multi-vendor dies and different process technologies.

Due to performance, density and reliability reasons, many current applications for microelectronic systems require the integration of sensing and actuation peripheral functions. The reuse of microelectronics technology, design and software tools makes it possible to integrate micro-electro-mechanical structures and electronics into a single *microsystem* chip. For example, microsystems can be fabricated on existing production lines for integrated circuits, with an additional post-processing step referred to as micromachining.

As chip designers face increasingly time-to-market constraints, any form of design reuse becomes essential to meet higher levels of integration. *Cores* may represent a possible choice for optimized designs, since they can associate some degree of customization to design reuse.

Depending on the application, performance requirements, time-to-market constraints, etc, state-of-the-art integrated systems may be designed as multi-chip modules, as microsystems or using embedded cores. Obviously, a lot of validation and testing challenges come along with these complex systems. The specific testing problems associated to them are addressed in the following sections.

4.1   Multi-chip Modules

Although multi-chip modules (MCMs) have been widely used for some years now, one of the most challenging problems in today's designs is still to meet manufacturing quality requirements. To ensure manufacturing quality, and hence acceptable assembly yields, high fault coverage tests are needed, that check the substrate, the bare dies and the module assembly.

Traditionally, the detection of faults in unpopulated substrates can be achieved by means of mechanical or electron-beam probing (Zorian, 1992). Regardless the technique in use, the goal of *substrate testing* is to detect open circuits, by checking the continuity of interconnects, and short circuits, by searching for unexpected charged interconnects.

Bare die testing is another crucial issue in MCM testing. Low speed functional and parametric tests are traditionally applied at the wafer level, in order to qualify dies according to their I/O driving capabilities, leakages, etc. But, while these tests suffice for dies which will be next encapsulated, dies to use in MCMs also require that at-speed performance tests and burn-in, usually achieved only at the packaged chip, are somehow performed at the wafer level. Performance characterization can be achieved by qualification of lots based on packaging samples. This is a cheap but imprecise procedure. Depending on the module assembly technology, burn-in can be carried out at the die level. If burn-in has to be postponed to the module verification step, either the testing costs will increase due to rework, or the final module yields will decrease. All these reasons make it very hard for chip suppliers to provide *known good dies* to MCM manufacturers.

The *final module testing* aims basically at checking the module assembly and its performance specifications. Then, interconnects must be verified, in order to detect mechanical defects due to low quality bonding or wrong die orientation. Next, dies need to be retested, since damages may occur due to the handling and bonding processes. Finally, the module must be tested for propagation times, that include dies and substrate routing delays. In case of fault detection, the module may undergo a repair process, consisting of procedures for faulty part isolation, rework and retest. Considering all these testing phases, and in order to keep fault detection and diagnosis tractable for complex MCMs, suppliers unanimously recommend that design-for-test techniques are applied to dies and module substrates.

Ideally, all dies would embed *built-in self-test* and *boundary scan* (Zorian, 1992). Built-in self-test would make it possible at-speed testing and burn-in of bare dies, and functional chip testing after die assembly. Boundary scan would help in checking the die assembly, in testing the interconnects and in verifying the performance of the assembled module. Additionally, it would be used for isolating faulty parts.

Even though the MCM market evolves very rapidly, one cannot reasonably expect that full chip level built-in self-test and module level boundary scan will be immediately reached. In the meantime, we can figure out an intermediate stage where MCMs will be populated by dies designed with boundary scan (some of them possibly including built-in self-test) and dies without any testability feature. Then, one possibility is to take advantage of the existing dies with boundary scan, to also test and diagnose dies and interconnects that have no testing capabilities (Lubaszewski, 1994). The global efficiency of this method will depend on the percentage of dies with and on the complexity of the dies without testability features. Another possibility is to integrate into the substrate the mechanisms needed to test dies and interconnects (Maly, 1994). This technique is illustrated in figure 16 for the digital boundary scan capability. For the sake of simplicity, only the interconnects of the boundary scan serial path are shown in the figure. It can be noted that, additionally to the boundary scan cells, a test access port controller for the non-boundary scan dies is built into the module substrate. It is clear that, in order to ensure acceptable yield levels, the density of active components in the substrate must be kept low enough to allow for relaxed design rules. This method has been successfully applied to digital (Maly, 1994) and to mixed-signal (Kerkhoff, 1997) MCMs.

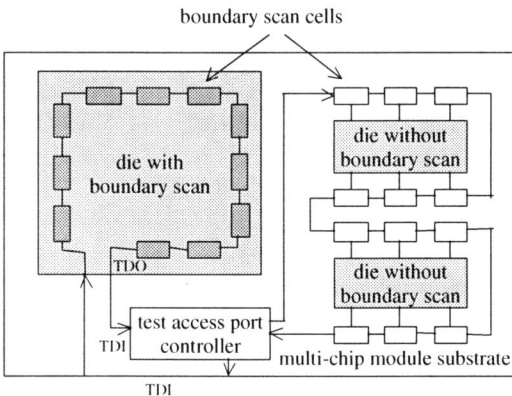

Figure 16. A multi-chip module smart substrate system

4.2   Microsystems

Similarly to the design and manufacture of MCMs, some existing microelectronics test techniques can be reused, and others need to be developed to cope with emerging *micro-electro-mechanical structures* (MEMS).

Up to date, few works have addressed some of the specific questions involved in testing the non-electrical parts of microsystems. (Vermeiren, 1996) has pointed out many problems related to modeling faults in MEMS, and it has discussed what would be requirements to achieve fault simulation and test generation in these microstructures. (Olbrich, 1996) addressed the design-for-test of non-electrical elements, highlighting related fault simulation issues. Both works have made an important contribution to the state-of-the-art, but none has clearly systematized its results, making it difficult to reuse their testing techniques in a straightforward way.

In (Lubaszewski, 1998b), this systematization is achieved by means of a computer-aided testing environment. In this environment, microstructure faults are modelled according to two complementary approaches. The first approach is based on modelling single non-electrical defects as analog faults (multiple and parametric, in general). This is possible because the behavioural modelling of non-electrical elements is usually based on a set of equations of the same nature as those modelling electrical elements. In this approach, defects are injected into the microstructure fault-free description by changing the nominal values of the components of the modelling circuit. In this case, the modified description is referred to as a *mutant*. For instance, considering the electro-thermal converter in Figure 17a, one type of microbridge break is modelled by changing the rthf and cthf values in the analog circuit of Figure 17b. The second fault modelling approach is based on the addition of new elements to a microstructure behavioural model, in order to properly represent the distributed effects of some failures associated to micromachining. The additional elements, also used for modelling bridging faults in digital circuits, are referred to as *saboteurs*.

Instead of developing a completely new tool, the injection of faults into the input description of a circuit simulator brings a pragmatic solution for the fault simulation of microsystems (Lubaszewski, 1998b). Therefore, fault injection can be achieved by simply modifying the fault-free microstructure description, and instantiating mutants and saboteurs from a *fault model library*. In addition to digital and analog single faults occurring in intrinsic (thermal, current, etc.) microsensors, this library must model multiple parametric, bridging and open circuit faults, missing and exceeding material defects, for a variety of MEMS.

Similarly to the scheme given in Figure 3, the test stimuli search process for the detection of faults in MEMS is assisted, in (Lubaszewski, 1998b), by a fault simulator. The test generation procedure is based on the sensitivity-guided process developed in (Nagi, 1993b) for analog circuits. On one hand, this test generation procedure cannot guarantee that it will find the best set of test stimuli, be the criterion the error maximization or the minimal number of test stimuli. On the other hand, it can manage mutants and saboteurs exactly in the same way as it manages conventional single faults. If, similarly to (Slamani, 1994), sensitivity figures were

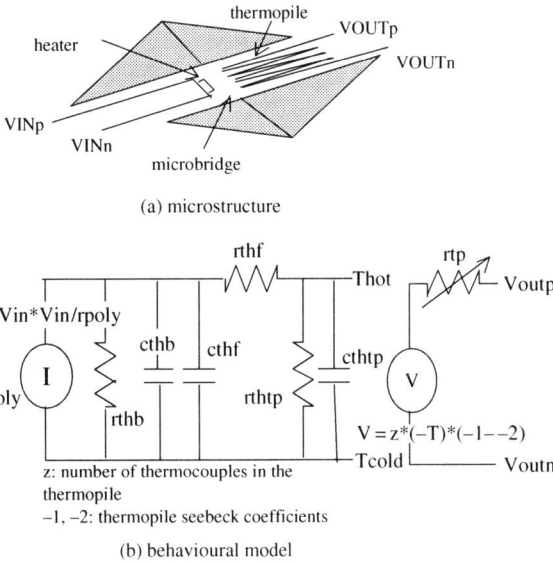

Figure 17. Electro-thermal converter

analytically computed, we might approach the best solution, but certainly could not avoid the exploding complexity of dealing with multiple analog faults (in the case of mutants), and increasingly complicate transfer functions (in the case of saboteurs).

Finally, the computer-aided testing environment of (Lubaszewski, 1998b) also provides a set of design-for-test strategies. These strategies range from ad hoc solutions derived from defect-oriented experiments, to the conversion of sensors into oscillators and the use of built-in current and thermal sensors.

4.3   Core-based Integrated Circuits

A wide range of embedded cores can be found in current systems-on-chip: memories, microprocessors, interfacing and networking cores, etc. These cores can be available in different forms: as a layout (hard core), as a netlist (firm core) or as a synthesizable register-transfer-level description (soft core). Although the test strategies for these cores may differ from one to another type, it is clear that, similarly to the motivations to go for core-based systems, the *reuse* of available testing techniques shall dominate the core supplier and system designer choices.

Ideally, the answer to the testing problems of core-based systems is to embed complex cores with *built-in self-test* capabilities accessed through a *boundary scan* infrastructure. In the case of low complexity cores, such as bus processing units, it may suffice just providing core access from outside the system. *Hard cores* that embed no testability at all may become easily testable, if they are surrounded by user-defined logic, specifically designed to give access, to provide test vectors and to collect core test responses. Then, the core supplier should provide the system designer with the tests needed for these cores to reach the desired fault coverage (Anderson, 1997).

As much as possible, built-in self-test might be shared with multiple cores, in order to avoid unnecessary redundancy and to lower the test hardware cost. However, it happens to hard cores, such as embedded RAMs, that system designers are not given the access to the core's built-in self-test. A solution is then to provide parameterizable items whose testability features can, not only change with the function, but also with the environment that will receive the core (Aitken, 1996). *Soft cores* make design and test more flexible, allowing for function programmability. In addition, if self-test is not built into a soft core, it will always be possible to synthesize the core using a commercial tool that automatically inserts internal and boundary scan. In this case, the fault coverage will depend on the core implementation given by the synthesis tool. In general, it may be expected that suppliers will soon provide cores packaged with *scan-based* and *built-in self-tests*, as the means to protect their *intellectual property* at the same time as keeping highly competitive.

5.   CONCLUSIONS

Existing design-for-test schemes and related test methods were extensively discussed in this chapter. The digital and analog cases were addressed, and testing issues were covered from the circuit to the system level. The major advantages of design-for-test over traditional test methods can be summarized as follows:
- the enhanced accessibility to internal test points makes it possible to develop short test sequences that achieve high fault coverage. This leads to high quality tests, requiring short application times. As a consequence, reliability is improved and time-to-market is shortened;
- cheaper testers can be used, as performance, interfacing and functional requirements are relaxed. Design-for-testability, built-in self-test and self-checking alleviate the probing requirements for the test equipment. Built-in self-test and self-checking alleviate functionality, since test generation and/or response evaluation are performed on-chip. In addition, the tester is no more required to operate at frequencies higher than that of the circuit under test, since built-in self-test allows for the application of autonomous at-speed testing.

The main drawbacks that come along with design-for-test are the following:
- an additional time is needed to design the test mechanisms to embed into integrated circuits and systems. However, the test development times for conventional testing methods are often longer. An alternative is to reuse pre-designed test cores;
- extra silicon is required to integrate test capabilities. However, embedded test structures have evolved over the years, and can now achieve very low area overheads. Additionally, the cost of transistors continues to drop;

– the performance of the circuit under test may be degraded by the additional test structures. Again, embedded test structures are expected to evolve, offering more efficient solutions. This already happens in the digital domain, but it is still a challenge for analog testing.

With the advent of submicron technologies, accessibility problems become worse due to many buried wiring layers. Additionally, higher operating frequencies bring ever more demanding requirements for the performance of testers. Therefore, design-for-test remains a good test solution for the emerging integrated systems. However, new testing developments are still needed. For example, fault models must be extended to include interconnect crosstalk. Also, robust designs for test structures are required, in order to get along with the reduced noise margins associated to low voltage circuits.

Reuse has been the keyword in the domain of integrated systems design. As new synthesis-for-test tools and test standards are developed, reuse tends also to dominate the testing of integrated circuits and systems. In fact, in the test domain this paradigm may not be limited to reuse pre-developed test cores in new designs. It can be further extended to reuse the same embedded test cores to perform different types of tests in different phases of a circuit's lifetime. These tests would allow for prototype debugging, post fabrication checking, maintenance testing, and concurrent error detection in the field. Only mechanisms based on unified off-line and on-line tests can add this dimension to the test reuse.

REFERENCES

Abramovici, M., Breuer, M.A. and Friedmann, A.D. (1990) Digital Systems Testing and Testable Design, Computer Science Press, New York.

Aitken, R.C., Butler, K.M., Campbell, R.L., Koenemann, B.K., Maly, W., Needham, W., Zorian, Y. and Roy, K. (1996) A D&T Roundtable: Deep-Submicron Test. *IEEE Design and Test of Computers*, fall, 102-8.

Akers, S.B. and Jansz, W. (1989) Store and Generate Built-In Testing Approach, in *International Test Conference*, Proceedings, 257-63.

Anderson, D.A. (1971) Design of Self-Checking Digital Networks using Coding Techniques, Technical Report 527, University of Illinois, Urbana, CSL.

Anderson, T.L., Chandramouli, R., Dey, S., Hemmady, S., Mallipeddi, C., Rajsuman, R., Walther, R., Zorian, Y. and Roy, K. (1997) A D&T Roundtable: Testing Embedded Cores. *IEEE Design and Test of Computers*, April-June, 81-9.

Argüelles, J., Martínez, M. and Bracho, S. (1994) Dynamic Idd Test Circuitry for Mixed-Signal ICs. *Electronics Letters*, **30**, number 6.

Armstrong, D.B. (1972) A Deductive Method of Simulating Faults in Logic Circuits. *IEEE Transactions on Computers*, **21**, number 5, 464-71.

Bardell, P.H., McAnney, W. and Savir, J. (1987) Built-In Test for VLSI: Pseudorandom Techniques, John Wiley and Sons, New York.

Benowitz, M. et al. (1975) An Advanced Fault Isolation System for Digital Logic. *IEEE Transactions on Computers*, **24**, 489-97.

Bracho, S., Martínez, M. and Argüelles, J. (1995) Current Test Methods in Mixed-Signal Circuits, in *Midwest Symposium on Circuits and Systems*, Proceedings, 1162-7.

Brackel, G.V., Glässer, U., Kerkhoff, H.G. and Vierhaus, H.T. (1995) Gate Delay Fault Test Generation for Non-Scan Circuits, in *European Design and Test Conference*, Proceedings, 308-12.

Bratt, A.H., Richardson, A.M.D., Harvey, R.J.A. and Dorey, A.P. (1995) A Design-for-Test Structure for Optimising Analogue and Mixed Signal IC Test, in *European Design and Test Conference*, Proceedings, 24-33.

Carter, W.C. and Schneider, P.R. (1968) Design of Dynamically Checked Computers, in *IFIP Congress*, Proceedings.

Caunegre, P. and Abraham, C. (1996) Fault Simulation for Mixed-Signal Systems. *KAP Journal of Electronic Testing: Theory and Applications*, **8**, 143-52.

Chatterjee, A. (1991) Concurrent Error Detection in Linear Analog and Switched-Capacitor State Variable Systems using Continuous Checksums, in *International Test Conference*, Proceedings, 582-91.

Cota, E.F., Lubaszewski, M. and Di Domênico, E.J. (1997) A New Frequency-domain Analog Test Generation Tool, in *International Conference on Very Large Scale Integration*, Proceedings, 503-14.

Courtois, B. (1981) Failure Mechanisms, Fault Hypotheses and Analytical Testing of LSI-NMOS (HMOS) Circuits, in *VLSI Conference*, Proceedings, 341-50.

Dandapani, R., Patel, J. and Abraham, J.A. (1984) Design of Test Pattern Generators for Built-In Self-Test, in *International Test Conference*, Proceedings, 315-9.

Devarayanadurg, G. and Soma, M. (1994) Analytical Fault Modelling and Static Test Generation for Analog ICs, in *International Conference on Computer-Aided Design*, Proceedings, 44-7.

Devarayanadurg, G. and Soma, M. (1995) Dynamic Test Signal Design for Analog ICs, in *International Conference on Computer-Aided Design*, Proceedings, 627-9.

Dufaza, C. and Cambon, G. (1991) LFSR based Deterministic and Pseudorandom Test Pattern Generator Structures, in *European Test Conference*, Proceedings, 27-34.

Eichelberger, E.B. and Williams, T.W. (1978) A Logic Design Structure for LSI Testability. *Journal of Design Automation and Fault-Tolerant Computing*, **2**, number 2, 165-78.

Francesconi, F., Liberali, V., Lubaszewski, M. and Mir, S. (1996) Design of High-Performance Band-Pass Sigma-Delta Modulator with Concurrent Error Detection, in *International Conference on Electronics, Circuits and Systems*, Proceedings, 1202-5.

Fujiwara, H. and Shimono, T. (1983) On the Acceleration of Test Generation Algorithms. *IEEE Transactions on Computers*, **32**, 1137-44.

Galay, J.A., Crouzet, Y. and Vergniault, M. (1980) Physical Versus Logical Fault Models MOS-LSI Circuits: Impact of their Testability. *IEEE Transactions on Computers*, **29**, number 6, 527-31.

Gardner, J.W. (1994) Microsensors, Principles and Applications, John Wiley and Sons, New York.

Goel, P. (1981) An Implicit Enumeration Algorithm to Generate Tests for Combinational Logic Circuits. *IEEE Transactions on Computers*, **30**, 215-22.

Gregorian, R. and Temes, G.C. (1986) Analog MOS Integrated Circuits for Signal Processing, John Wiley and Sons, New York.

Hayes, J.P. (1976) Transition Count Testing of Combinational Circuits. *IEEE Transactions on Computers*, **25**, 613-20.

Huertas, J.L., Vázquez, D. and Rueda, A. (1992) On-line Testing of Switched-Capacitor Filters, in *VLSI Test Symposium*, Proceedings, 102-6.

IEEE Standard 1149.1. (1990) IEEE Standard Test Access Port and Boundary Scan Architecture, IEEE Standards Board, New York.

Kerkhoff, H.G., Boom, G. and Speek, H. (1997) Boundary Scan Testing for Mixed-Signal MCMs, in *International Conference on Very Large Scale Integration*, Proceedings, 515-25.

Khaled, S., Kaminska, B., Courtois, B. and Lubaszewski, M. (1995) Frequency-based BIST for Analog Circuit Testing, in *VLSI Test Symposium*, Proceedings, 54-9.

Kolařík, V., Mir, S., Lubaszewski, M. and Courtois, B. (1995) Analogue Checkers with Absolute and Relative Tolerances. *IEEE Transactions on Computer-Aided Design of Integrated Circuits and Systems*, **14**, number 5, 607-12.

Koenemann, B., Mucha, J. and Zwiehoff, G. (1979) Built-in Logic Block Observation Techniques, in *Test Conference*, Proceedings, 37-41.

Krishnan, S., Sahli, S. and Wey, C.-L. (1992) Test Generation and Concurrent Error Detection in Current-Mode A/D Converter, in *International Test Conference*, Proceedings, 312-20.

LeBlanc, J.J. (1984) LOCST: A Built-In Self-Test Technique. *IEEE Design and Test of Computers*, November, 45-52.

Lo, J.C., Daly, J.C. and Nicolaidis, M. (1992) Design of Static CMOS Self-Checking Circuits using Built-In Current Sensing, in *International Symposium on Fault Tolerant Computing*, Proceedings, 104-11.

Lu, A.K. and Roberts, G.W. (1994) An Analog Multi-tone Signal Generator for Built-In Self-Test Applications, in *International Test Conference*, Proceedings, 650-9.

Lubaszewski, M. and Courtois, B. (1992) On the Design of Self-Checking Boundary Scannable Boards, in *International Test Conference*, Proceedings, 372-81.

Lubaszewski, M., Marzouki, M. and Touati, M.H. (1994) A Pragmatic Test and Diagnosis Methodology for Partially Testable MCMs, in *MultiChip Module Conference*, Proceedings, 108-13.

Lubaszewski, M., Mir, S., Rueda, A. and Huertas, J.L. (1995) Concurrent Error Detection in Analog and Mixed-Signal Integrated Circuits, in *Midwest Symposium on Circuits and Systems*, Proceedings, 1151-6.

Lubaszewski, M., Mir, S. and Pulz, L. (1996) ABILBO: Analog BuILt-in Block Observer, in *International Conference on Computer-Aided Design*, Proceedings, 600-3.

Lubaszewski, M. and Courtois, B. (1998a) A Proposal of Reliable Fail-Safe System. *IEEE Transactions on Computers*, **47**, number 2, 236-41.

Lubaszewski, M., Cota, E.F. and Courtois, B. (1998b) Microsystems Testing: an Approach and Open Problems, in *Design, Automation and Test in Europe*, Proceedings, 524-8.

Maly, W. and Nigh, P. (1988) Built-In Current Testing – Feasibility Study, in *International Conference on Computer-Aided Design*, Proceedings, 340-3.

Maly, W., Feltham, D.B.I, Gattiker, A.E., Hobaugh, M.D., Backus, K. and Thomas, M.E. (1994) Smart-Substrate Multichip-Module Systems. *IEEE Design and Test of Computers*, summer, 64-73.

Marchok, T.E., El-Maleh, A., Maly, W. and Rajski, J. (1995) Complexity of Sequential ATPG, in *European Design and Test Conference*, Proceedings, 252-61.

Maunder, C.M. and Tulloss, R.E. (1990) The Test Access Port and Boundary Scan Architecture, IEEE Computer Society Press.

Maunder, C.M. (1994) The Test Access Port and Boundary Scan Architecture: An Introduction to ANSI/IEEE Std. 1149.1 and its Applications, in *Forum on Boundary Scan for Digital and Mixed-Signal Boards*, CERN, Geneva.

Meixner, A. and Maly, W. (1991) Fault Modelling for the Testing of Mixed Integrated Circuits, in *International Test Conference*, Proceedings, 564-72.

Milor, L. and Visvanathan, V. (1989) Detection of Catastrophic Faults in Analog Integrated Circuits. *IEEE Transactions on Computer-Aided Design*, **8**, number 2, 114-30.

Mir, S., Lubaszewski, M., Liberali, V. and Courtois, B. (1995) Built-In Self-Test Approaches for Analogue and Mixed-Signal Integrated Circuits, in *Midwest Symposium on Circuits and Systems*, Proceedings, 1145-50.

Mir, S., Lubaszewski, M. and Courtois, B. (1996a) Fault-based ATPG for Linear Analog Circuits with Minimal Size Multifrequency Test Sets. *KAP Journal of Electronic Testing: Theory and Applications*, **9**, 43-57.

Mir, S., Lubaszewski, M., Kolarík, V. and Courtois, B. (1996b) Fault-based Testing and Diagnosis of Balanced Filters. *KAP Journal on Analog Integrated Circuits and Signal Processing*, **11**, 5-19.

Mir, S., Lubaszewski, M. and Courtois, B. (1996c) Unified Built-In Self-Test for Fully Differential Analog Circuits. *KAP Journal of Electronic Testing: Theory and Applications*, **9**, 135-51.

Mir, S., Rueda, A., Olbrich, T., Peralías, E. and Huertas, J.L. (1997) SWITTEST: Automatic Switch-Level Fault Simulation and Test Evaluation of Switched-Capacitor Systems, in *Design Automation Conference*, Proceedings.

Murray, B.T. and Hayes, J.P. (1996) Testing ICs: Getting to the Core of the Problem. *IEEE Computer*, November, 32-8.

Nagi, N., Chatterjee, A. and Abraham, J.A. (1993a) DRAFTS: Discretized Analog Circuit Fault Simulator, in *Design Automation Conference*, Proceedings, 509-14.

Nagi, N., Chatterjee, A., Balivada, A. and Abraham, J.A. (1993b) Fault-based Automatic Test Generator for Linear Analog Circuits, in *International Conference on Computer-Aided Design*, Proceedings, 88-91.

Nagi, N., Chatterjee, A. and Abraham, J.A. (1994) A Signature Analyzer for Analog and Mixed-Signal Circuits, in *International Conference on Computer Design*, Proceedings, 284-7.

Nagi, N., Chatterjee, A., Balivada, A. and Abraham, J.A. (1995) Efficient Multisine Testing of Analog Circuits, in *International Conference on VLSI Design*, Proceedings, 234-8.

Nicolaidis, M. and Courtois, B. (1988a) Strongly Code Disjoint Checkers. *IEEE Transactions on Computers*, **37**, number 6, 751-6.

Nicolaidis, M. (1988b) A Unified Built-In Self-Test Scheme: UBIST, in *International Symposium on Fault Tolerant Computing*, Proceedings, 157-63.

Nicolaidis, M. (1990) Efficient UBIST Implementation for Microprocessor Sequencing Parts, in *International Test Conference*, Proceedings, 316-26.

Nicolaidis, M. and Boudjit, M. (1991) New Implementations, Tools and Experiments for Decreasing Self-Checking PLAs Area Overhead, in *International Conference on Computer Design*, Proceedings, 275-81.

Nicolaidis, M. (1993a) Efficient Implementations of Self-Checking Adders and ALUs, in *International Symposium on Fault Tolerant Computing*, Proceedings, 586-95.

Nicolaidis, M. (1993b) Finitely Self-Checking Circuits and their Application on Current Sensors, in *VLSI Test Symposium*, Proceedings, 66-9.

Nicolaidis, M. (1994) Efficient UBIST for RAMs, in *VLSI Test Symposium*, Proceedings, 158-66.

Nigh, P. and Maly, W. (1989) Test Generation for Current Testing, in *International Test Conference*, Proceedings, 194-200.

Ohletz, M. (1991) Hybrid Built-In Self-Test (HBIST) for Mixed Analog/Digital Integrated Circuits, in *European Test Conference*, Proceedings, 307-16.

Olbrich, T., Richardson, A., Straube, B. and Vermeiren, W. (1996) Integrated Test Support for Micro-Electro-Mechanical-Systems (MEMS), in *International Conference on Micro Electro, Opto Mechanical Systems and Components*, Proceedings, 273-9.

Osseiran, A. (1995) Getting to a Test Standard for Mixed-Signal Boards, in *Midwest Symposium on Circuits and Systems*, Proceedings, 1157-61.

Rajski, J. and Tyszer, J. (1992) The Analysis of Digital Integrators for Test Response Compaction. *IEEE Transactions on Circuits and Systems II*, May, 293-301.

Rajsuman, R. (1992) Digital Hardware Testing: Transistor-level Fault Modelling and Testing, Artech House Inc.

Renovell, M., Lubaszewski, M., Mir, S., Azais, F. and Bertrand, Y. (1997) A Multi-Mode Signature Analyzer for Analog and Mixed Circuits, in *International Conference on Very Large Scale Integration*, Proceedings, 65-76.

Roth, J.P., Bouricius, W.G. and Schneider, P.R. (1967) Programmed Algorithms to Compute Test to Detect and Distinguish between Failures in Logic Circuits. *IEEE Transactions on Electronic Computers*, **16**, number 10, 567-80.

Savir, J. (1980) Syndrome-Testable Design of Combinational Circuits. *IEEE Transactions on Computers*, **29**, number 6, 442-51.

Sebeke, C., Teixeira, J.P. and Ohletz, M.J. (1995) Automatic Fault Extraction and Simulation of Layout Realistic Faults for Integrated Analogue Circuits, in *European Design and Test Conference*, Proceedings, 464-8.

Seshu, S. (1965) On an Improved Diagnosis Program. *IEEE Transactions on Electronic Computers*, **12**, number 2, 76-9.

Slamani, M. and Kaminska, B. (1993) T-BIST: A Built-In Self-Test for Analog Circuits based on Parameter Translation, in *Asian Test Symposium*, Proceedings, 172-7.

Slamani, M. and Kaminska, B. (1994) Multifrequency Testability Analysis for Analog Circuits, in *VLSI Test Symposium*, Proceedings, 54-9.

Slamani, M. and Kaminska, B. (1995) Multifrequency Analysis of Faults in Analog Circuits. *IEEE Design and Test of Computers*, **12**, number 2, 70-80.

Smith, J.E. and Metze, G. (1978) Strongly Fault Secure Logic Networks. *IEEE Transactions on Computers*, **27**, number 6, 491-9.

Soma, M. (1990) A Desing-for-Test Methodology for Active Analog Filters, in *International Test Conference*, Proceedings, 183-92.

Soma, M. and Kolarík, V. (1994) A Desing-for-Test Technique for Switched-Capacitor Filters, in *VLSI Test Symposium*, Proceedings, 42-7.

Szekély, V., Marta, Cs., Rencz, M., Benedek, Zs. and Courtois, B. (1995) Design for Thermal Testability (DFTT) and a CMOS Realization, in *Therminic Workshop*.

Szekély, V., Rencz, M., Karam, J.M., Lubaszewski, M. and Courtois, B. (1998). Thermal Monitoring of Self-Checking Systems. *KAP Journal of Electronic Testing: Theory and Applications*, **12**, 81-92.

Tsai, S.J. (1991) Test Vector Generation for Linear Analog Devices, in *International Test Conference*, Proceedings, 592-7.

Toner, M.F. and Roberts, G.W. (1993) A BIST Scheme for an SNR Test of a Sigma-Delta ADC, in *International Test Conference*, Proceedings, 805-14.

Ulrich, E.G. and Baker, T.G. (1974) Concurrent Simulation of Nearly Identical Digital Networks. *IEEE Computer*, **7**, number 4, 39-44.

Vermeiren, W., Straube, B. and Holubek, A. (1996) Defect-Oriented Experiments in Fault Modeling and Fault Simulation of Microsystem Components, in *European Design and Test Conference*, Proceedings, 522-7.

Wadsack, R.L. (1978) Fault Modelling and Logic Simulation of CMOS and MOS Integrated Circuits. *The Bell System Technical Journal*.

Zorian, Y. (1992) A Universal Testability Strategy for Multi-Chip Modules based on BIST and Boundary Scan, in *International Conference on Computer Design*, Proceedings, 59-66.

CHAPTER 9

SYNTHESIS OF FPGAs AND TESTABLE ASICs

DON W. BOULDIN

Electrical Engineering, 1508 Middle Drive, University of Tennessee, Knoxville, TN 37996-2100 U.S.A.,
Tel: (865)-974-5444, Fax: (865)-974-5483, E-mail: dbouldin@tennessee.edu

Abstract: Industrial designers and educators who plan to design microelectronic systems (e.g. hardware accelerators, co-processors, etc.) are increasingly capturing their designs using hardware description languages such as VHDL and Verilog. The designs are then most often synthesized into programmable logic components such as field-programmable gate arrays (FPGAs) offered by Xilinx, Altera, Actel and others. This approach places the emphasis on high-level design which reduces time to market by relying on synthesis software and programmable logic to produce working prototypes rapidly. These prototypes may then be altered as requirements change or converted into high-volume mask gate arrays or other application-specific integrated circuits (ASICs) when the demand is known to be sufficient. These ASICs, however, must be designed to be testable to screen out those with manufacturing defects. Hence, scan logic must be inserted, test vectors generated and fault grading performed to ensure a high level of testability. These efforts complicate and delay the conversion of FPGA designs to ASICs but must be considered by designers of microelectronic systems. Topics covered include: design flow; system partitioning; hardware description languages (HDLs); specifying behavioral control; specifying structural components; critical paths; placement and routing; technology choices; FPGA applications; rapid prototyping; retargeting; manufacturing defects; scan chain insertion; test vector generation; fault grading, and ASIC production

Keywords: VHDL, FPGA, ASIC, synthesis, programmable logic, testing

1.   INTRODUCTION

Designing microelectronic systems involves mapping application requirements into specifications that can then be implemented using appropriate microelectronic components. These specifications must be represented at every level of abstraction including the system, behavior, structure, physical and process levels. Internal functions must be described as well as the interactions among these components and the external world.

R. Reis et al. (eds.), Design of Systems on a Chip, 191–201.
© 2006 *Springer.*

Some of the distinguishing characteristics (Bouldin, 1991) of microelectronic systems are:
- specified hierarchically,
- conform to an interface specification,
- incorporate computing (analog and/or digital processing),
- constructed using microelectronic components, and
- involve input/output devices (e.g. sensors and actuators).
Some example microelectronic systems are:
- a portable instrument for monitoring environmental data,
- an accelerator coprocessor board in a personal computer,
- a controller for a robot or an automobile, and
- a data compressor for transmitting facsimiles or images.

This chapter presents an overview of the methodology for developing microelectronic systems. Thus, the role of hardware description languages, synthesis, physical placement and routing software, programmable logic, testing and microelectronic components will be delineated. Several caveats to this methodology in terms of price and performance will also be discussed.

2.   MICROELECTRONIC COMPONENTS

The designer of a microelectronic system frequently employs several existing integrated circuits (ICs) to meet the requirements of an application and thus adds value to the final product by interconnecting components in a unique way. Increasingly, software is also added to provide flexibility that further distinguishes the designer's product from those produced by competitors. Use of existing components reduces the time required to implement the design and generally leads to higher profits. The components are relatively inexpensive since they are commodity parts produced in high volume (millions of copies). Microprocessors and other VLSI/LSI components such as digital signal processing chips are frequently the most cost-effective since thousands of gates costing only tens of dollars (or millions of gates costing only hundreds of dollars) can be utilized. SSI and MSI components are appropriate for tasks which involve mostly external communication and very little internal processing. Packaging is the dominant factor influencing the cost of these components. Hence, SSI and MSI components are rarely the best choice for implementing logic functions requiring hundreds of gates. Analog components such as operational amplifiers, analog-to-digital converters and digital-to-analog converters are utilized to interface to sensors and actuators.

Off-the-shelf components are by necessity general-purpose so optimum performance and/or cost for a specific application may not be achieved. However, a variety of user-specified components or application-specific integrated circuits (ASICs) can be developed by the designer if the situation warrants. In these cases, the designer is able to implement only those functions needed for a special-purpose application so that very little of the physical space is wasted. Thus, the production cost per integrated circuit is held to a minimum.

Techniques have been developed to permit the designer to use a programmable power supply to specify one or two of the layers in some semicustom integrated circuits. These field-programmable gate arrays (FPGAs) contain fewer functions than mask gate arrays whose layers are specified using optical masks since space is required for the programmable links. FPGAs are also slower because of the increased resistance and capacitance of the links. However, the time required for customization is only a few minutes or hours as compared to several weeks for the mask gate arrays or ASICs (Trimberger, 1994).

FPGAs are presently used to implement logic functions up to 200,000 gates or more with a production quantity of 200,000 or less. Mask gate arrays (MGAs) or standard-cell ASICs are used for designs requiring more gates, higher speed or higher volume production. In one style of MGAs, the vendor prefabricates rows of gates with spaces or channels between the rows allocated for interconnections. In another style, the chip appears as a sea of gates (actually transistors) in which some of the gates are used for processing and others for interconnections. The first style is used for less complex designs since the physical space is used less efficiently. Both styles have been adopted to make the task of performing automatic placement and routing straightforward. Mask gate arrays can be fabricated in only 3-5 weeks since the vendor stockpiles wafers and needs only to have the interconnection masks made and the wafers processed for the remaining layers.

FPGAs that are equivalent in number of gates to MGAs may cost 2-10 times as much. The additional silicon area or processing consumed for the programming elements and the on-board addressing circuitry make the fabrication of FPGAs much more expensive. On the positive side, extensive testing can be performed by the supplier prior to delivery to the designer, reducing some product development costs.

3.   PRODUCT DEVELOPMENT

The development of a product requires careful consideration of many factors including the requirements of the application, the availability of appropriate micro-electronic components, familiarity with electronic design automation tools and the experience of the designers. Perhaps equally significant are other factors such as the perceived demand in the marketplace, the market window (period in which the product is expected to sell), the presence of competition, and the risk of developing and introducing new technologies, These factors combine to place a tremendous pressure on developers to accelerate the design and prototyping stages of product development in an effort to reduce the time-to-market to a minimum. Studies (Huber and Rosneck, 1991) have shown that for every month's delay in being introduced to the market, a ten percent decrease in profits is experienced. Thus, those in market-ing must proceed swiftly to ascertain the functionality desired by the customer and those in design must quickly capture these in hardware. The designer must use his time wisely and cleverly but not overlook errors (such as the now infamous division error detected in the Pentium chip (Wirbel, 1994). Whenever these processes are performed too quickly, there is an increased risk of errors in the design, or just as unfortunate, an increased risk that what is produced will not satisfy the customer's needs.

Time-to-market pressures force designers to select off-the-shelf or semicustom components as described above. It is generally very beneficial to produce a hardware prototype as soon as possible since extensive verification of the design cannot be determined without it. Statistics (Huber and Rosneck, 1991) have shown that for mask gate array designs only one set of masks is required in order to obtain prototypes which are fully functional in a stand-alone mode. This mode consists of testing the single integrated circuit on a tester using the same vectors that were applied to the software simulator that modeled the IC. In essence, electronic design automation tools have matured sufficiently that this first-pass success has become routine for digital circuitry. However, these same statistics have shown that these working prototypes fail about half the time when placed in the final system when they are subjected to inputs and outputs from other components. This failure has been attributed to insufficient system-level simulation or modeling. While simulating, designers are just not subjecting the new ASIC to a realistic view of its ultimate environment. Even if a high degree of realism can be achieved, the simulation may consume an overwhelming amount of resources (computer time, memory and disk space). In some cases, no adequate model even exists. For example, image manipulation circuitry must be presented to the human for evaluation and our understanding of the human visual system is still quite primitive. In other cases, analog circuitry is involved and is not modeled with sufficient precision.

This situation has provided great impetus to programmable logic since a design can be implemented temporarily using FPGAs and then later retargeted to mask gate arrays. Thus, the designer can practice using programmable logic and embed the hardware in the full system environment to perform verification to ensure that he has captured the design correctly. Since the penalty for making an error is quite small at this point compared to having to endure the expense for additional mask and wafer processing, designers can rush through the software simulation. Although some design errors may not be caught until testing the hardware, at least these tests can be applied and evaluated quickly, usually at full system speed. These hardware prototypes can also be shipped to potential customers for beta testing. In fact, the prototypes can be considered as first-release production products which can be updated later with a new part or, in the case of reconfigurable components, with a new configuration file. The developers are thus given the opportunity to be more certain that what is about to be produced in quantity will satisfy the customers. Therefore, the risks involved are reduced significantly.

Conversion from FPGAs to mask gate arrays is generally performed in order to obtain a more cost-effective solution if the demand for the product is sufficient. To calculate when this cross-over point occurs, the designer must evaluate not only the costs of the FPGA parts versus the mask gate array parts but also the number needed and the cost of conversion.

Table 1. Comparison of product development

Design Stage	FPGA (weeks)	ASIC (weeks)
Design Specification	1.0	1.0
Design Entry	1.6	1.6
Functional Simulation	2.4	4.0
Test Vector Generation	0.0	6.4
Vendor Interface	0.0	1.6
Prototype Test	1.6	1.6
Prototype Lead Time	0.0	2.0
Production Lead Time	0.0	6.0
Total Design Cycle	7.0	24.0

Table 2. Comparison of product costs

Expense	FPGA (dollars/part)	ASIC (dollars/part)
Raw unprogrammed part	8.00	4.00
Design/Simulation	3.15	7.92
Manuf. Test Vectors	0	2.88
Place/Route/Masks	0	2.20
Final Part	11.15	17.00

Tables 1 and 2 (after Lytle, 1997) compare the product development cycles and the major costs of the two technologies. In both cases, 20,000 copies of the part must be produced and each part must contain 20,000 gates. It is assumed that the slower speed of the FPGA part is acceptable. It should be noted that in 2006 these numbers are more likely to be 200,000 gates and 200,000 copies but the procedure for performing the calculations here has not changed.

The initial difference between the two technologies is the additional time required for simulation. Because the penalty for making an error using a mask gate array is several thousand dollars and a schedule slip would further damage time-to-market and anticipated profits, the designer is likely to spend almost twice as long simulating the mask gate array.

The next significant difference in developing these technologies is the need to generate manufacturing test vectors for the mask gate array. Unquestionably, this stretches the development time and adds to the overall cost, as shown in Table 2. Even though the FPGA initially costs twice as much as the mask gate array, the additional expenses for generating manufacturing test vectors and for the vendor's one-time manufacturing charges make the final mask gate array cost more than the final programmed FPGA. Thus, for this example, the crossover point for converting to the mask gate array is greater than 20,000 parts. This number has risen by a factor of two in just the past two years because of rapidly falling prices for FPGAs (Lytle, 1997).

4.   DESIGN METHODOLOGY

Figure 1 illustrates the prevalent design methodology for semicustom components. The designer begins by interpreting the application requirements into architectural specifications which can be implemented in one or more microelectronic technologies. It is not likely that this step will be automated since it involves mapping abstract concepts (often described in narrative form) into precise statements which depend on the capabilities of available microelectronic components. This task is extremely

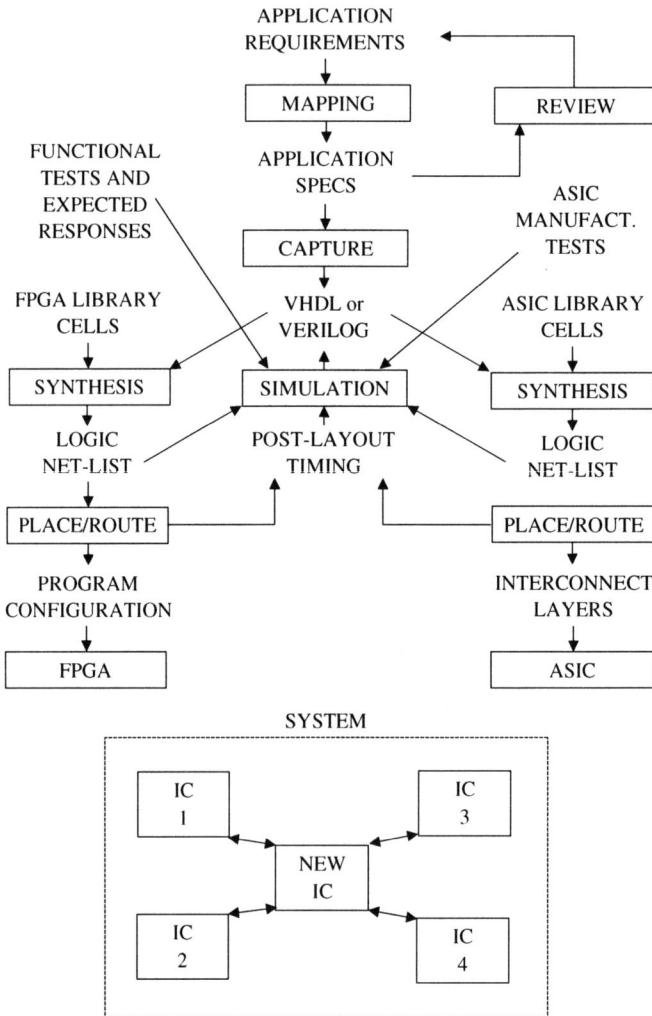

Figure 1. Design of a microelectronic system for multiple technologies

complex and the quality of the resulting implementation is greatly impacted by the experience and creativity of the designer.

4.1   Hardware Description Languages and Synthesis

Once the application requirements are understood, the designer translates the architectural specifications into behavior and/or structure representations. Behavior denotes the functionality required as well as the ordering of operations and completion of tasks in specified times. A structural description consists of a set of components and their interconnection. These components may be primitives or collections of primitives. Both behavior and structure may be specified using hardware description languages such as Verilog or VHDL (Very High Speed Integrated Circuit Hardware Description Language). These text-based languages permit complex hierarchies to be managed efficiently and may even be required for large designs consisting of thousands of logic gates. HDLs can be translated automatically into net-lists of library components using synthesis software. This software performs essentially three functions: (1) translation from text to a Boolean mathematical representation, (2) optimization of this representation based on one or more criteria such as size or delay or testability, and (3) mapping or binding of the optimized mathematical representation to a technology-specific library of components.

HDLs appear initially like other software languages and deceive many designers into thinking that writing code in an HDL is just like writing other software code. However, these languages are tailored for describing hardware and thus permit concurrent operations. A traditional electrical engineer expects hardware components to be active simultaneously and HDLs permit this situation to be modeled. However, traditional computer programmers expect a single CPU that performs operations sequentially. These programmers have expressed dismay at trying to program multiprocessors operating in parallel, but that is exactly why HDLs are used. Thus, a designer who visualizes the hardware will write code which is more efficiently manipulated by the synthesis software. However, he should avoid micro-managing the hardware description too much or otherwise there is nothing left for the synthesis tool to do.

The use of synthesis benefits the designer in several ways. First, it enables him to capture the design in a straightforward manner that may more closely parallel the same way in which the designer envisioned the tasks. This is especially true for describing the behavior of a controller or finite state machine since often the designer is thinking in terms of a collection of if-then-else processes. The text is easily modified in many cases and facilitates the management of large designs. This is not to say, however, that text should be used exclusively. Graphical schematics are often superior at expressing the interconnection of components. Fortunately, electronic design automation tools have been developed which permit graphical schematics to be produced from textual representations and vice-versa. Thus, a description can even be a mixture of text and graphics so that the designer can use whichever representation is warranted.

Another benefit of synthesis is improvement in the designer's productivity. It takes only a few minutes for the synthesis tool to perform a variety of optimizations on the captured design so the designer does not have to spend hours or perhaps days looking for redundancies or trying to minimize delays. The synthesis tool can be invoked multiple times to provide several candidate solutions from which the designer can select at his leisure. This process essentially trades computer cycles for human sweat. It has been reported that a novice designer using synthesis can obtain a solution in a few days which approximates the same quality as that which an experienced designer might obtain in several weeks. However, synthesis should not be considered a panacea since inefficient mapping can result in larger and slower designs than those produced by humans.

In addition to improved productivity, another benefit of using synthesis is the ability to retarget the design without having to recapture it. Figure 1 illustrates that the design can be captured once and the synthesis tool invoked more than once with different technology libraries. Obviously this is of great benefit when programmable logic is used to practice for the final design that is implemented for cost considerations using mask gate arrays. However, this approach can be helpful whenever the designer decides to switch families of programmable logic. For example, a design which takes several months to develop can immediately take advantage of new product offerings. This approach can also be used to select the most cost-effective part for a particular application since it is possible to evaluate several technologies before purchasing the best one. Similarly, the designer may find it necessary to switch to a second source for mask gate arrays and retargeting can make this easy. Just having the retargeting capability in hand maintains a competitive environment for the hardware suppliers that can impact both their price and service.

For those designers who believe they can produce higher quality designs manually, there is still a role for synthesis. The software can be used to obtain an adequate solution quickly and then the designer can pinpoint those portions which need his expert attention. Thus, he can use his time wisely and enhance an otherwise poor solution.

4.2   Physical Placement and Routing

Once the synthesis tool produces a net-list of technology-specific library components the design is ready to be placed and routed. The effort required for designs of even a few thousand gates is prohibitive if performed entirely by a human. It is much more efficient to invoke placement and routing software for this task and to intervene only in rare cases in which the software does not find a solution for some nets. The designer can also accept a partial solution and micro-manage the placement and routing of only those critical nets which restrict operating the system clock at a higher frequency. The complexity of the placement and routing task is so great that software cannot be used to obtain an exact solution but instead must be invoked iteratively in a clever manner. The prevailing algorithm in use today is based on simulated annealing since this optimization routine is tuned to obtain a global optimum rather than accept some local optimum and stop searching the solution space.

Placement and routing is highly order dependent in that the options remaining for a net are reduced each time another has been routed. Consequently, nets designated by the designer to be critical are routed first. This ensures that these nets will be given preferrential treatment and are highly likely to be acceptable. However, some of the remaining nets may not be treated so kindly and more intervention will be required. Fortunately, acceptable solutions can be obtained quickly or the designer can elect to use a larger part which has more logic and routing resources. This adds expense but at least the design gets implemented.

Programmable logic differs from mask parts in that the programming elements introduce a significant RC delay which makes FPGAs perhaps five or ten times slower than MGAs which use metal vias and lines for interconnection. To counter this problem to some degree, FPGA suppliers use segments of varying length to avoid stepping into too many *puddles*. This process also adds complexity to the optimization routines in the placement and routing software.

5.   PRODUCING TESTABLE ASICs

FPGA prototypes may then be altered as requirements change or converted into high-volume mask gate arrays or other application-specific integrated circuits (ASICs) when the demand is known to be sufficient. These ASICs, however, must be designed to be testable to screen out those with manufacturing defects. Hence, scan logic must be inserted, test vectors generated and fault grading performed to ensure a high level of testability. These efforts complicate and delay the conversion of FPGA designs to ASICs but must be considered by designers of microelectronic systems.

5.1   Testing Requirements

Test stimuli which validate the desired functional behavior of a circuit provide for functional testing only. Every design must be subjected to these tests to ensure that the application requirements are met by the circuit that is under construction. In essence, if the application requires an adder function then the circuit being designed must be checked to demonstrate that addition is being performed and not subtraction or some other unwanted function.

Functional test stimuli are first used inside a simulator to validate the desired behavior. They are then applied to the first hardware prototypes to be sure that the desired circuit has been manufactured. Once a *golden* copy of the circuit has been produced in hardware, additional copies of the part are tested to be certain they are true replicas of the golden one. This additional testing of the production copies is termed *manufacturing testing*. Manufacturing tests are required since the semiconductor manufacturing process can contain numerous defects which in turn lead to yields of less than 100%. In fact, yields of 20-50% are not uncommon with new processes (Weste and Eshraghian, 1993).

To screen out defective circuits, several steps are taken. First, a calibration coupon is placed on each wafer in 4-6 locations. These are probed to determine whether the calibration circuits behave within acceptable tolerance on the wafer in question. Next, power is applied to individual dies on the wafer and the circuits are probed to determine if the quiescent current is within an acceptable range. Those circuits not passing this test are marked with an ink dot and discarded. Those passing are packaged for subsequent, more extensive tests.

Packaged integrated circuits are generally placed first on a stand-alone tester and subjected to a variety of test stimuli at a relatively low speed of 1 Mhz. Those which pass are then subjected to higher speed tests and possibly burn-in or environmental tests. ICs which pass all of these tests are shipped to the designer who then inserts each IC into the target system and runs complete system-level, at-speed tests to be certain that only working parts are sent to the customer.

5.2   Faults, ATPG and Scannable Logic

Manufacturing defects may manifest themselves in a variety of ways including shorts and opens which make the logic nodes appear to be stuck at one or stuck at zero. Some defects give rise to the desired behavior but only after unacceptable delays. Fortunately, these faults can be modeled and graded so that only those parts which exhibit an acceptable quality level (generally above 98%) will be sent to the customer. Also, numerous techniques have been developed for automatic test pattern generation (ATPG) so that these goals can be achieved.

It is not uncommon to apply ATPG software to a circuit only to learn that a portion of the circuit is either uncontrollable or unobservable from the primary inputs/outputs. One countermeasure is to insert a probe point in the circuitry which can be accessed externally. Perhaps the simplest means of implementing these probes is to connect external terminals to existing internal storage elements. If the circuit is still not controllable or observable, additional storage elements may be inserted or the circuit rearranged. This activity is termed *design-for-testability*.

One means of accessing internal probes while minimizing the space consumed for interconnections is to connect them in serial fashion. This produces a chain of flip-flops which enable external data to be scanned in and responses to be scanned out. Since the flip-flops must be able to support normal circuit operation as well testing, each one must be preceded by a multiplexer. These devices are therefore known as *scannable* flip-flops. This additional wiring and multiplexing adds slightly to the overall cost of the circuitry but must be included or else faulty parts will be sent to the customer.

The use of scannable flip-flops at the periphery of the circuit is termed *boundary scan*. Having access at these points is sufficient to determine whether a detected fault lies within the part or in one of its neighbors. *Built-in self-test (BIST)* refers to circuitry within the component which generates or applies test stimuli to the remaining internal circuitry. One obvious advantage of BIST is the fact that the tests are applied at full-speed. Another is the ease of access to internal functions and an almost unlimited number of inputs and outputs. Yet another benefit of including BIST is that the test may be initiated at any time, even after being shipped to the customer. Thus, the customer can reinitiate the BIST and be reassured whether the part is working faithfully.

Both FPGAs and ASICs must be testing for manufacturing defects. In the case of FPGAs, the vendor can perform these tests prior to shipping them to the application designer. Hence, the application designer needs only to generate functional tests and check that the programming of each part is a faithful reproduction of his prototype. However, the application designer who uses ASICs must shoulder the additional burden of generating and applying warranted manufacturing tests. Thus, it is the responsibility of the application designer to use scannable flip-flops, generate the manufacturing test patterns and perform fault grading. If the fault grade is too low, the designer must insert additional scannable flip-flops or redesign the circuitry. All of this must be done prior to manufacturing in order to be certain that the desired level of fault grading can be achieved. After the parts have been fabricated, each one must be subjected to these manufacturing tests to screen out the faulty ones.

6.   SUMMARY

Designers of microelectronic systems are increasingly capturing their designs using hardware description languages such as VHDL and Verilog. Designs requiring up to 200,000 gates and less than 200,000 copies are most often synthesized into FPGAs. When higher performance or larger quantities are warranted, these designs are retargeted to ASICs. However, ASICs must be designed to be testable to screen out those with manufacturing defects. Hence, scan logic must be inserted, test vectors generated and fault grading performed to ensure a high level of testability.

REFERENCES

Bouldin, D., Borriello, G., Cain, T., Carter, H., Kedem, G., Rabaey, J. and Rappaport, A. (1991) *Report of the Workshop on Microelectronic Systems Education in the 1990's*, Knoxville, TN: Electrical Engineering, University of Tennessee.

Huber, J. and M. Rosneck, M. (1991), *Successful ASIC Design the First Time Through*, New York: Van Nostrand Reinhold.

Lytle, C., (1997) *The Altera Advantage: First Quarter Newsletter*, p. 3.

Trimberger, S. (ed.) (1994) *Field Programmable Gate Array Technology*, Norwell, MA: Kluwer Academic Publishers.

Weste, N. and Eshraghian, K. (1993) *Principles of CMOS VLSI Design (2nd Ed.)*, Reading, MA: Addison-Wesley Publishing.

Wirbel, L. (1994) IBM Stops Shipping Pentium-Based PCs, *Electrical Engineering Times*, Issue **828**, p. 14 (December 19, 1994).

CHAPTER 10

TESTABLE DESIGN AND TESTING OF MICROSYSTEMS

HANS G. KERKHOFF

MESA+ Research Institute/University of Twente, Testable Design & Testing of Microsystems Group, 7500 AE Enschede, The Netherlands, e-mail H.G.Kerkhoff@el.utwente.nl

Abstract: The low internal node accessibility of microsystems and the variety of components and technologies used in these systems can be considered as a real challenge for testing these devices. Although many concepts can be applied which have already been developed for testing boards and ICs, also new aspects like sensor/actuator testing have to be considered. The choice which of the many approaches is most adequate turns out to be very application dependent. Test approaches are shown for two microsystems

Keywords: Design-for-Test, Testing, BIST, Test Busses, Boundary-Scan, Sensor/Actuator Testing, Microsystems, MCMs

1.   INTRODUCTION

The developments in miniaturisation of complete systems are going very fast. Often this provides economic realisations of products with new features not envisioned before. We will emphasise on microsystems consisting of (a) sensor part(s), analogue and digital signal processing and (an) actuator part(s). A brief introduction in the technology and applications of microsystems has been presented by Wilkinson in this book. Microsystems usually involve many different implementation technologies. Hence these systems are often built up of different parts and mounted all together on a single substrate. Possible implementation forms are hybrids or Multi-Chip Modules (MCMs).

The testing of these complex systems goes beyond conventional electrical testing, as also embedded sensors and actuators are involved. The requirement of testing these microsystems is very application dependent. Consumer microsystems consisting of few parts require cheap testing approaches with no need for diagnosis, while complex military microsystems require extensive testing and often diagnosis

R. Reis et al. (eds.), Design of Systems on a Chip, 203–218.
© 2006 *Springer.*

is needed for repair of the microsystem after production. In the case of safety-critical applications, also issues like self testing become of interest.

A direct result of the miniaturisation of systems is a dramatically decreased accessibility, and hence decreased testability of these systems. The testing of these microsystems can include digital hardware-software testing, analogue/mixed-signal testing and sensor/actuator testing. Based on developments in ICs and PCBs, there are a number of ways of making these microsystems testable, although many new test problems emerge.

With regard to testing the digital parts, already much work has been carried out which can be directly used in microsystems. Digital multiplexing, digital busses, Built-In Self-Testing (BIST) of regular structures and a variety of (boundary) scan techniques have been developed in the past. In the digital signal processing chips boundary-scan cells can be integrated. Another approach even implements this hardware in the (active) silicon substrate of the MCM.

For analogue and mixed-signal parts the situation is much more complex. One approach makes use of existing data converters in the microsystem and the flexibility of digital signal processing techniques. Also concepts like (mixed-signal) boundary-scan can be applied, while other approaches like analogue busses or BIST techniques are being investigated at this moment.

The front- and back-end of microsystems are very difficult to test. This is because often non-electrical stimuli are required or non-electrical responses have to be measured. Sometimes, the required packaging for handling the physical quantities involved in the microsystem (e.g. a fluid) complicate the testing of the system even more. In several cases where these physical properties can be electrically generated/converted on chip or substrate there are possibilities for self testing. Examples of devices where electrical generation is potentially feasible are magnetic and optical devices.

Two examples of sensor-based microsystems will be presented and the testing approach we used will be shown. The first is an electronic compass wrist watch employing magnetic flux-gate sensors and the second is a Micro Analysis System (MAS) for evaluating reaction products resulting from mixed fluids.

2.   MICROSYSTEMS AND THEIR DIFFERENT APPLICATIONS

2.1   Introduction

With regard to testable design and testing of microsystems, the final application area is of extreme importance. The application area relates to the direct and indirect costs involved in testable design and testing. Indirect costs can e.g. come from damage claims in safety-critical applications or overall loss of e.g. a space-borne system in a professional application. We will distinguish between high-volume consumer (low-cost) applications, low-volume professional (medium- to high-cost) applications and safety-critical (high-cost) applications.

2.2 Microsystems for High-Volume Consumer Applications

If one looks at high-volume consumer applications, such as e.g. hand-held audio equipment (CD-, mini-disc player), video equipment (hand-held TV), computers (PC, Personal Digital Assistant) and telecommunication (GSM) apparatus, then already many of these applications can be considered as microsystems. Even regular electronic equipment is often constructed by means of microsystems.

The small sensors used can be of any kind, such as optical, audio, magnetic and electromagnetic. The same holds for the actuators (e.g. optical, audio).

In this application area, there is very little room for testing and Design-for-Test hardware (DfT), as these costs come back in every of the great many devices. In addition, the competition is fierce and the consumer very cost-driven. Usually, testing is confined to automated application tests, where the features of the sensors, actuators and signal processing in combination are tested.

2.3 Microsystems for Low-Volume Professional Applications

The second mostly used application areas of microsystems are the ones used in industry for professional use. Examples are measurement units for measuring different physical quantities in industrial process control. In the latter case, e.g. sensors are extended with micro-electronics to carry out preconditioning calculations and the implementation of (standardised) communication with a central computer. From a testing point of view this is quite favourable, as a natural modular set-up of the total system is guaranteed. Sometimes the requirements are such that some form of (measurement) stability of the systems has to be guaranteed, in which case auto-calibration of these "smart sensors" comes in the picture. Actually, the latter can be considered as some form of self testing capability.

In professional micro systems, some possibilities are open for DfT and testing beyond application tests. As an example we will show the Micro Analysis System, (MAS) which is able to find different properties of a fluid.

2.4 Microsystems for Safety-Critical Applications

Probably the most interesting application area from a scientific point of view is safety-critical systems. Examples are medical systems (e.g. pacemakers, bionic control) or transportation (car, plane) systems control (e.g. ABS, airbag, automatic collision control). In these applications periodic or on-line tests have to be carried out of all parts to ensure the proper operation or, besides indication of an error, activation of a redundant microsystem. It is obvious that in these systems DfT and testing have the greatest potential at this moment. Although not always safety-critical but certainly asking for similar measures are very expensive systems (e.g. a communication satellite) which heavily depend on certain critical microsystem parts.

3. PROBLEMS AND SOLUTIONS INVOLVED IN TESTING MICROSYSTEMS

3.1 Introduction

The essential test problems in microsystems can be summarised as inaccessibility, high complexity and many different technologies. In a modern microsystem, often parts of the signal processing are based on embedded software.

This means that hardware/software co-design testing is required. Unfortunately, at this moment still much basic research has to be carried out in this area.

Digital testing has matured and many good approaches and tools are available nowadays, but high-speed testing and testing asynchronous designs still require future attention.

Analogue and mixed-signal testing are still not at the same level of development as compared to digital testing. An analogue fault-model is not yet accepted, and fault simulation and test-pattern generation tools still have a long way to go. Much research effort will be required in this area.

Testing sensors and actuators on board is still quite a virgin area, especially in generating and evaluating non-electrical quantities. With regard to fault modelling, fault simulation and test-signal generation similar problems are present as in the analogue case; however, the situation is even worse because of the gap between electrical and non-electrical quantities in terms of available tools.

3.2 Hardware/Software Codesign Testing

The set-up of many microsystems consists of sensors, actuators, analogue conditioning circuitry (amplification, filtering, data conversion) and digital signal processing. The latter will be often carried out by conventional processor cores, which enable embedded core testing, or specially designed coprocessors (e.g. for high-speed calculation of angles) which operate in conjunction with conventional processors. In the case of full design freedom, the choice can be made which parts of the calculations are carried out in hardware and which in software. In the most promising approach, test requirements are already included at the specification level (Vranken, 1994).

Examples of translating these requirements in the actual implementation are hardware Design-for-Test structures (e.g. scan, BIST) and software test routines. Software testing is basically a design verification problem. Problems in e.g. storage of software are considered to be hardware problems. Basically, structured programming, formal specification and reusability of (proven) software can help in creating bug-free software. As exhaustive software testing is non-practical, other methods have been developed, like unit testing (testing of a coded module against its module specification) and integration testing (verifying the communication between verified units). Other approaches use e.g. watchdog functions (results have to be within a certain number of calculation cycles or values). The area of hardware/software testing is still in its very infancy, but will undoubtedly be of much importance in future (flexible) microsystems.

3.3 Design-for-Test and Testing of the Digital Parts of a Microsystem

Digital testing has already a long history in terms of research and development (Abramovici, 1990). Since the beginning of the seventies, when complexities increased, it started to become a **research** area instead of being just a **service** activity. In the past decades, many test approaches have been developed as well as dedicated test hardware and test tools. Results of these developments can be found in proceedings of e.g. the International Test Conference, European Test Conference and VLSI Test Symposium. Within the scope of this paper, an extensive listing of many of the results is not relevant.

With regard to microsystems, the scan testing (Weyerer, 1992) approach seems to be the most promising. In the case of hybrids/MCMs, employing many different digital circuits of probably several manufacturers, the boundary-scan approach (Bleeker, 1992) provides a standard for a protocol for communication and control of the different tests. A further advantage is the possibility for (structural) automatic test-pattern generation of the system and advanced tools for diagnostics. A more recent extension of digital testing is the application of measuring quiescent currents in the power (Iddq) (Rajsuman, 1995). Also in the area of Built-In Self Test (BIST), already many approaches exist for regular structures such as memories or (co)processors, employing Linear Feedback Shift-Registers (LFSR) for data generation and evaluation. The availability of much processing power and memory on future microsystems will be able to contribute to self-testable microsystems.

3.4 Design-for-Test and Testing of the Analogue Parts of a Microsystem

The testable design and testing of analogue parts in microsystems lags seriously behind with regard to their digital counterparts. Usually, only (limited) functional tests are carried out. This results from the absence of any simple analogue fault models, and the still relatively unstructured design of analogue circuits. The introduction of defect-based testing, hierarchical fault simulators and analogue macro/module generators start to change these unfavourable starting points. The most promising approach assumes that, like in digital logic, analogue building blocks (macros) can be distinguished, such as amplifiers, filters and data converters. As testing approaches of these stand-alone blocks (e.g. data converters) are well known, the problem of accessibility remains. This problem is much more severe than in the digital world. A promising approach is the introduction of a new mixed-signal boundary-scan standard (1149.4) (Wilkens, 1997). As many important parameters of analogue macros can be found in the frequency domain, and these tests are often time/cost consuming, tests are carried out in the time domain and mathematically converted (DFT, FFT) in the frequency domain by means of digital signal processing (DSP). As many microsystems will have computing power aboard, this will present new possibilities for Built-In Self Tests based on DSP techniques. New developments in the area of analogue and mixed-signal testing can for instance be found in the proceedings of the International Mixed-Signal Testing Workshops.

3.4.1   The macro concept

Originally presented for digital circuits, the macro concept (Beenker, 1986) also has many advantages in the analogue/mixed-signal testing environment. Basically, a system is assumed to be build up from functional entities such as amplifiers, filters, data converters etc. In this case, the testing problem is reduced to testing the separate parts and their interconnections. Very often, the required tests for the macros have been well developed. These macros can also take the form of e.g. a bare die or a packaged SMT device; in essence the approach can be used at different levels of hierarchy. A problem encountered often in analogue systems parts, is that malfunctioning of preceding or succeeding macros is also able to deteriorate the behaviour of the (correct) macro in between.

3.4.2   How to obtain access to macros

There are several ways in which access (between microsystem inputs and local inputs, and microsystem outputs and local outputs) can be obtained (Kerkhoff, 1995). We will confine ourselves to electronic access, as this is the usual approach in real microsystems.

The most elegant way is to ensure that preceding and succeeding macros are transparent to the signals required for testing the macro in between. Unfortunately this can only be accomplished in some cases; furthermore it is often unknown in advance which macros will precede or succeed in a system application.

Another approach is to implement two high-quality switches at the input and output of a macro which are able to switch between the signal path and test lines/busses. The best approach is to isolate the macro completely from preceding and succeeding macros to avoid the previously mentioned loading problem. Alternatively, the output can be monitored at the input of the succeeding macro, thereby reducing negative effects on the signal path. It is obvious that these switches should be very good in not disturbing the original signals, and accurately (little loss or noise addition) connecting the signals to e.g. test busses. Single MOS transistors and transmission gates are often capable or realising this.

A more structured approach can make use of the mixed-signal boundary-scan standard (1149.4) (Wilkins, 1997). It is basically a multiplexer-based approach. An equivalent of the digital scan technique in analogue circuits seems not to be practical. Probably the most elegant way is to avoid any access and generate/evaluate signals locally. This is discussed in one of the next sections.

3.4.3   Data converter testing

As previously mentioned, many macro types have been evaluated in the past; hence the ways how to test or characterise these macros is quite well known. A very specific category of macros, the data converters (DAC & ADC) will be frequently found at the inputs or outputs of microsystems (Van der Plassche, 1994). In the past, many different tests have been developed for this category of macros and will therefore be briefly discussed as an example of macro testing.

Starting simple, for instance DC measurements can be carried out (e.g. with the help of an on-board DAC while testing an ADC), providing zero offset data, full-scale accuracy, gain errors, integral and differential non-linearity (INL & DNL) data. Dynamic tests, either using external or internal signals, can also reveal high bit-error rates, total harmonic distortion (THD), Signal-to-Noise Ratio (SNR) and dynamic INL & DNL values. A DSP-based approach makes use of histogram tests, looking at codes, and revealing INL & DNL, and missing codes. Which approach is most feasible depends on the application; the availability of processing power can make e.g. DSP tests economically feasible.

3.4.4   Digital Signal Processing (DSP)

Many interesting parameters in analogue circuits (and probably also sensors) have to be derived in the frequency domain. To generate and evaluate signals in the frequency domain is usually (test)time consuming and complex. In the case of self tests it is very difficult. To carry out measurements and evaluate signals in the time domain is easier. Some mathematical calculations like the Discrete Fourier Transformation (DFT) and Fast Fourier Transformation (FFT) are capable of transforming timing data into frequency data (Mahoney, 1987). From this, many parameters of macros can be calculated. With the increasing computing power, this method of testing and self-testing will become feasible for microsystems.

3.4.5   Built-In Self-Tests (BIST)

In some cases, like safety-critical applications (e.g. an insulin automatic injection system), regular (in-field) checks have to be carried out to verify the correct operation of a microsystem. In digital systems, BIST is not uncommon anymore for regular structures. In analogue systems, however, it is still very rare. One example (Ohletz, 1991), uses a DAC for the generation of several different DC voltages, assuming on-board ADC and DACs in most microsystems. After application of this signal to the analogue parts, the results are converted into digital data by means of an ADC after which further evaluation is carried out by pure digital (BIST) hardware. The actual generation/evaluation is basically a digital BIST structure which is also assumed to be available in most future microsystems. A recent development is evaluating the correctness of an analogue macro by means of bringing it into oscillation during the test mode. Any deviation of the correct operation will result in different oscillation frequencies (Arabi, 1996).

It must be concluded that the current approaches are not very effective yet, which leaves much room for further research.

3.5   The Testing of Sensors & Actuators

Most sensors in their most elementary form measure an analogue physical quantity (e.g. light, magnetic field, pressure) and convert this into an analogue voltage or current. In a similar way, actuators convert an analogue voltage or current into a physical analogue quantity. Because we assume them to be part of a microsystem, they are assumed to be small, with relatively small voltages and currents involved.

With regard to measuring or applying the analogue voltages or currents, they can be considered as **analogue macros**, able to employ the same DfT and test techniques as discussed in the previous section. Access can be obtained via analogue multiplexers and busses. Also the developed P1149.4 standard could contribute in this area. Sometimes, the values of the generated voltages and currents are very small and/or noisy. In such cases it makes sense to consider part of the analogue conditioning and control electronics as being part of the sensor/actuator. In the case of modular microsystems, employing a bus or protocol (IEEE 1451.3) this is the most likely approach.

For the sake of simplicity we will confine ourselves to sensors; in the case of actuators we have the dual case. First, very little is known about defect mechanisms in sensors. In addition, no tools are available for evaluation, except the ones developed for analogue circuits. As a result, structural test-generation is extremely rare. Experiments have started to introduce defects in sensors in an HDL-A description language, trying to connect it with the electronic parts in order to predict the behaviour of the total microsystem. Being a virgin area, still much work is required here. Until then, functional-based testing is the only way to approach the problem.

In a conventional test-bench environment, usually the required physical quantities (light, pressure etc.) can be generated, although it would be appreciated if one could do without it. It is however a requirement in safety-critical applications. The major problem in the self-testing of these sensor devices lies in generating the non-electrical physical quantities for which they are sensitive.

Basically we can distinguish between two categories:
1. Physical quantities which can be generated within the system by means of microsystem hardware. Some examples of these quantities are, in ranking order of complexity: temperature, magnetic fields, light and pressure. In all cases, it is assumed that the values of the quantities are relatively small.
2. Those quantities which cannot, or with great difficulty, be generated by means of microsystem hardware. Some examples are: fluid and gas properties (e.g. Lactose or CO sensitive) and extreme values of the quantities mentioned in 1.

It is obvious that the second category is very difficult to implement in the case of self testing. More realistic at this moment is to look at item 1. Temperature can be created by controlled dissipation of electrical power, magnetic fields can be generated by (planar) coils, light by means of LED structures, and pressure by piezo-electric devices. As additional option, the currents and voltages of the device in the absence of the quantity can be stored and compared during testing. In the case of self-testing, both the generation and evaluation should take place in the electrical domain. For instance, the compass watch could use additional coils for testing the sensor/actuator **combination**.

4.   EXAMPLES OF TEST APPROACHES OF MICROSYSTEMS

4.1   The Compass Wrist Watch

As a first example of a microsystem and its used test approach we will discuss the compass wrist watch (Tangelder, 1997). The global set-up of the system is shown in Figure 1. The system consists of two orthogonally placed planar magnetic flux sensors. In order to operate these sensors, triangular currents have to be applied. Any external change in the magnetic field will result in a time shift of the small output voltages. This data is converted into pulses and subsequently the compass function is calculated digitally. Our implementation uses a silicon MCM substrate, which contains a number of chips (mixed-signal circuits, a digital chip), some passive components (R, C) and two planar flux-gate sensors. The interconnections are made with high-density IC-lithographic techniques. The silicon substrate is able to provide means for active Design-for-Test hardware.

The tests to be carried out can be divided into several categories. As the implementation is based on bare dies/sensors on a (silicon) substrate (MCM), the substrate has been tested first for interconnection faults and correctness of the active and passive devices (Known Good Substrate, KGS). Resistivity-based tests take care for detecting opens and shorts in interconnections. In the case of active devices, conventional IC testing techniques can be applied. The bare dies and sensors have been tested functionally before assembly (Known Good Dies, KGD). The compass watch had the advantage that all chips have been designed in-house, thus enabling the introduction of scan chains and analogue access and test hardware in the chips. Often this is not the case in microsystems. As one of several alternatives, the active substrate could be used to introduce DfT structures.

After assembly, the digital parts have been evaluated. Structural (topology-based) test-pattern generation has been used, and the digital parts are equipped with conventional scan chains. Four different scan chains have been used in this design.

A mixed-signal verification test system has been used during evaluation of the microsystem.

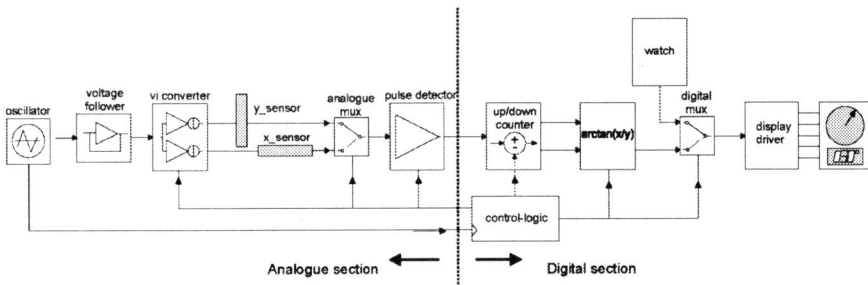

Figure 1. The Compass Watch microsystem set-up

   The analogue macros in this example, are an operational amplifier (connected as voltage follower), VI converters, triangle oscillator and a comparator (Figure 1). The macros are mostly preceded and succeeded by digitally controlled analogue multiplexers. Dynamic current testing has also been incorporated to verify some analogue macros (Bracho, 1997). This required the design of a current sensor. As an example, Figure 2 shows the set-up to test the operational amplifier. The switches have been implemented by MOS transistors. Digital control in the microsystem is accomplished via "test_a_b" and "test_b_c". The lines labelled "in" and "out" are the output and input of respectively the preceding and succeeding macros (triangular oscillator and VI converters). The remaining wires are connected to the analogue (microsystem) test busses on the substrate. Three tests are carried out for this operational amplifier to determine offset and frequency (amplitude & phase) behaviour.

   Figure 3 shows the measurement result of the frequency behaviour of this operational amplifier using the above DfT structures; the graphical representation is a virtual test instrument using the LABVIEW software package. Using the same approach, also the other analogue macros have been functionally verified. By combinations of macros, also the interconnections between the macros have been tested.

   The testing of the sensors can be accomplished in several ways. The flux-gate sensor for detecting the earth magnetic field is shown in Figure 4 (Kawahito, 1995). The charm is its completely planar construction. It consists of a permalloy core sandwiched between first and second-level metal. Effectively an excitation coil and a pickup coil is created, coupled by the magnetic layer.

   While functionally testing, the (external) test environment has to provide a specific magnetic field, which is not always easy in a (noisy) test environment. A more advanced approach, includes (e.g. 2) small planar coils in the vicinity of the sensor to create the required range in magnetic fields. The control electronics can be provided by the external tester or on-board electronics. Basically, the *combination* (sensor, actuator, electronics) is tested.

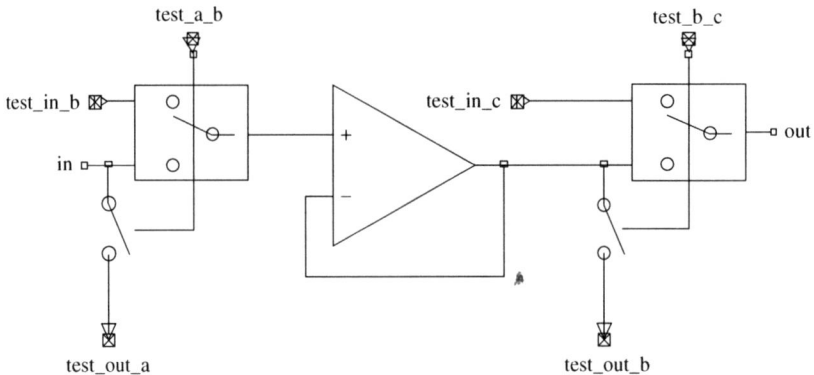

Figure 2. Example of added analogue DfT structures to isolate and test an operational amplifier

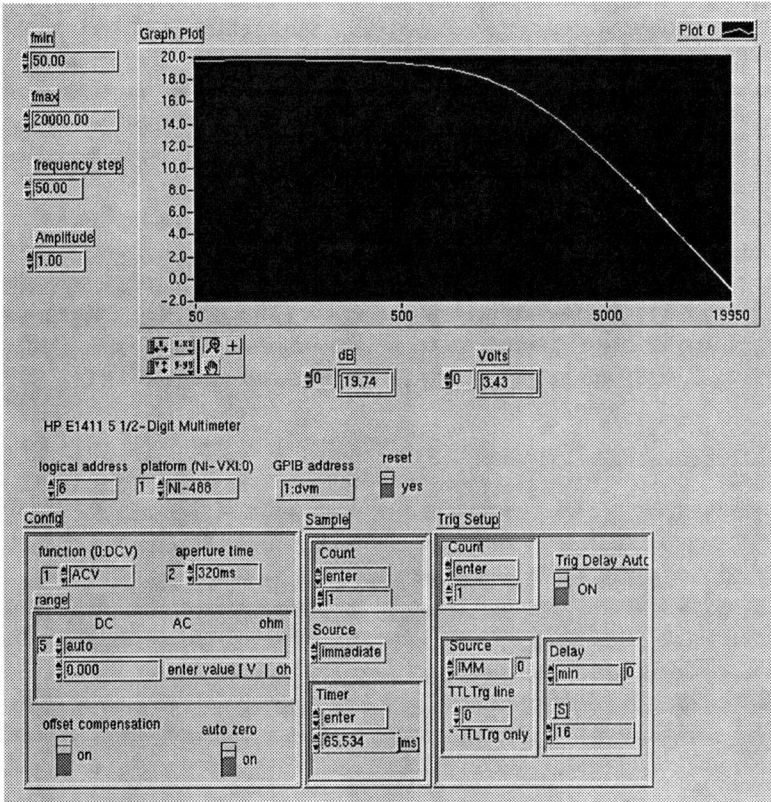

Figure 3. Virtual test instrument representation of the frequency test of the operational amplifier

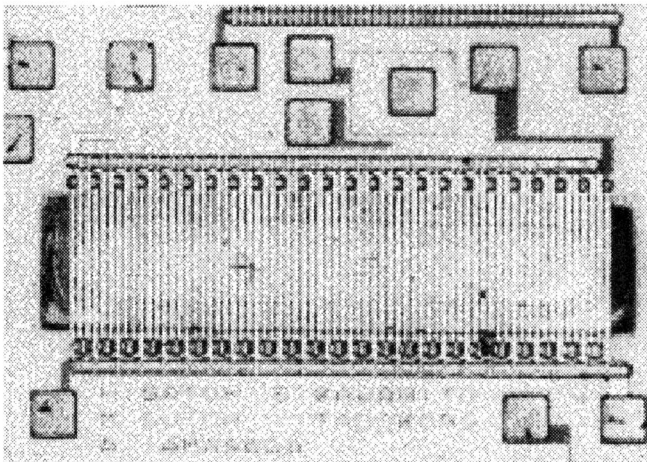

Figure 4. Microphotograph of a planar flux-gate sensor

The generated voltages of the flux-gate sensors are quite small and somewhat noisy. Hence, although analogue multiplexers have been implemented between the sensors and analogue conditioning circuitry, it is not wise to directly measure the voltages via the multiplexers, test busses and buffers. As the analogue parts have already been evaluated, the sensor signals are preferably measured after some internal amplification and inherent filtering. It also ensures the interconnections between sensors and analogue circuitry.

Finally, an overall functional application test is carried out. In the case of the compass watch, the previously discussed extensive test approach is only carried out during evaluation/characterisation and first prototype runs. Being a low-cost consumer product, only the application test is carried out after assembly. Diagnostics and repair is not an issue. Random taken samples undergo incoming inspection (functional) tests verify the bare dies, sensors/actuators and substrate.

4.2   The Micro Analysis System

As a second example of a microsystem and its proposed test approach, the Micro Analysis System (MAS) is presented (Van den Berg, 1996). It consists of three micro fluid channels etched in a silicon substrate, two inlets and one outlet, with a flow micro-sensor and a micropump in each of the inlets, and an optical absorption module at the outlet. The overall scheme is shown in Figure 5. The system basically measures chemical reaction products from two fluids (sample and reagent) by detection of the spectral absorption intensity. The microelectronics is located on top of the silicon substrate. A microphotograph of the flow sensor and a fluid channel

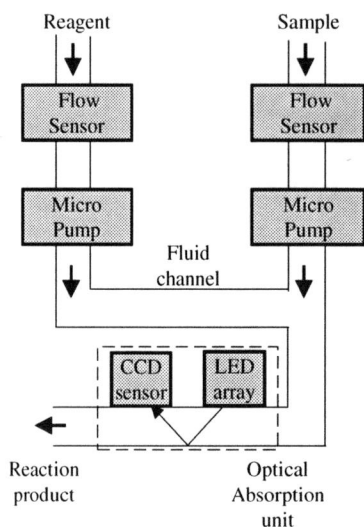

Figure 5. Basic scheme of a Micro Analysis System (MAS)

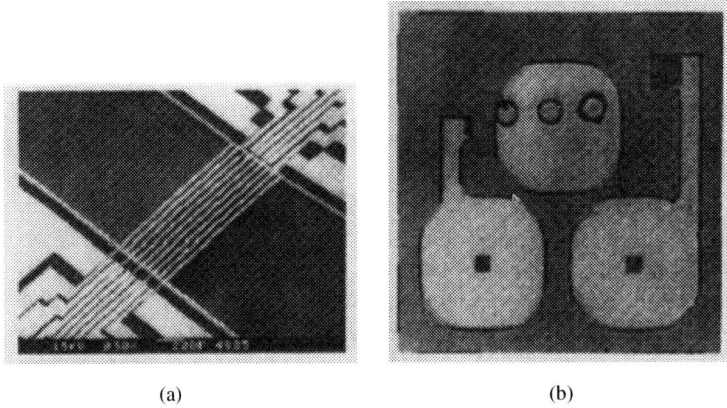

(a)   (b)

Figure 6. (a) The resistor-based flow sensor with fluid channel in the silicon substrate (b) The silicon, resistor-based micro pump with two valves

is seen in Figure 6a and the micropump in Figure 6b. The optical absorption sensor consists of the combination of a small optical sensor (e.g. Charge-Coupled Device (CCD)) and an optical source (e.g. LED/laser array), as shown in Figure 5. In order to have both devices on top of the channel, a reflector in the channel is required.

The testing approach used, especially for the sensors/actuators, is directly dependent on the actual microsystem implementation. In this example, the fluid channels are etched in the silicon substrate (Figure 6a) and the two resistance-based flow sensors are made in the substrate too (Lammerink, 1995). All other devices can be glued on top of the substrate later on, and interconnected via wire bonding with the interconnection patterns on the substrate. The silicon-based micropump (Figure 6b) is also using resistors, and is glued on the silicon substrate too; this is a high-risk manufacturing step as the mechanical valves have to work properly at the glued silicon-silicon interface.

This microsystem implementation can also use the previously discussed known good die/actuator/sensor and known good substrate test principle. In this particular case, however, the "known good substrate" principle has a different meaning than in the previous example. This has to do with the method of manufacturing the flow sensors and micropumps. Although the testing of the sensors and actuators is basically carried out in a functional way, now new test problems arise with respect to the flow sensors and the micropumps as these devices cannot be considered to be *substrate-independent* devices. This means that these devices cannot be manufactured independently from the substrate in contrast to the independently manufactured LED array and CCD devices and additional digital and analogue integrated circuits. Furthermore, the substrate incorporates fluid micro channels which can be jammed, making the substrate useless.

Hence, the very first test to be carried out is verifying the substrate/flow-sensors combination. In the case of a passive substrate (only interconnection wires and passive devices) one may require substrate interconnection tests, and resistance tests for the electrical integrity of the resistor-based flow-sensors and passive devices. Capacitive tests can verify the correct operation of integrated capacitors.

The second step is the assembly of the micropumps on the substrate and again resistive testing of the electrical integrity of the resistor-based micropumps can be carried out. The next step has to be necessarily a full functional test, including fluid interaction. This is required with regard to the correct micropump valve operation, and unobstructed passage in the fluid channels which are *pure mechanical* operations. This requires testing the flow-sensor/micropump combinations, which can be subsequently converted into electrical signals which can be verified by a tester system. The tester in fact now emulates part of the system in this approach.

The following step is the assembly of the optical absorption unit (Figure 5). Before assembly, these bare-die devices have been functionally verified separately. The LED/laser array and CCD can be tested purely functional and are handled as bare dies. Light emission and associated currents are verified in the LED array. The CCD is tested as digital shift-register and as light/dark sensor. As repair of these devices after assembly is also rather difficult, because of the transparent layer and the transparency deterioration after repair of this layer, also a combined functional test is carried out afterwards. The following sensor/actuator test involved is verifying the optical absorption module. The CCD shift operation is tested by pure electronic methods. The LED array is not powered during this operation. Next, the LED array is sequentially activated and the response of the CCD compared digitally. The LED array activation can be monitored separately by measuring the power current (Figure 7).

As all other dies are mounted on this combined substrate later on, with the option of relative easy repair, the correct operation and testing of the sensors/actuators/substrate combination (including electrical and non-electrical properties) is vital. The following step is the assembly of all digital and analogue bare dies.

The digital chip, consisting of the clock generator, pulse generator, control and DSP block is tested using structural ATPG and (boundary) scan techniques ((B)S) as well as functional tests (Figure 7). As this chip is locally designed, the boundary-scan

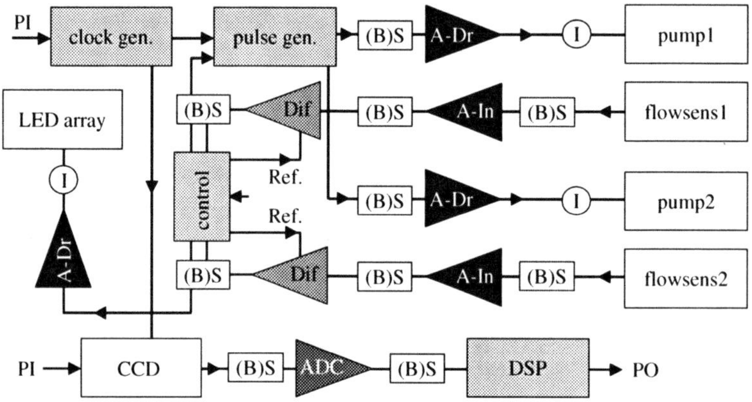

Figure 7. Possible architecture of MAS and an extensive use of DfT structures

cells are incorporated in the chip. Structural ATPG-based tests as well as functional tests can be carried out at the bare die level.

There are two analogue/mixed-signal chips. One incorporates the Amplifying Drivers (A-Dr) and Integrators (A-In) and differential amplifiers (Dif). This locally designed chip is equipped with mixed-signal boundary-scan cells (Wilkins, 1997) and current sensors (Figure 7). The analogue-to-digital converter (ADC) is bought as bare die, and typically not equipped with any DfT. A possibility is to incorporate mixed-signal boundary-scan cells in the active substrate, located around this bare die (Kerkhoff, 1997). The bare dies can be tested with standard analogue functional tests and standard ADC tests.

As Figure 7 shows, all interconnections between the chips, including wire-bondings after assembly, can be tested using the boundary-scan cells. The same holds for the structural integrity of the bare dies on the substrate. A selection of previous bare-die tests can be chosen subsequently for structural or functional tests. Any repair on chips and/or wire-bondings should be completed now. Hence, the testing of the total system behaviour remains.

The parallel set-up of the inlets of this microsystem (Figure 5) is used for testing after complete assembly, using fluids. First the left branch is verified/activated and next the right branch. The serial connection of the flow sensor and micropump, and the (correct) optical absorption module, can determine whether the flow sensor or the micropump or both are malfunctioning. These tests verify the interconnections between the sensors/actuators and the electronics, and unobstructed passage in the fluid channels which is basically a *strictly mechanical process*.

If the module response differs from the (standard) reagent fluid value, the micropump is not correct (assuming reagent fluid at the inputs). If this is not the case, and the flow sensor does not provide any signal, then the flow sensor is not correct. As actuators often consume power (like a micropump), an absence of any current during operation can also be used for diagnostics. These tests are repeated for the other branch. The control electronics of the sensors and actuators has already been verified in the early stage. As already stated, any malfunctioning now of the sensors/actuators and channels will result in rejection of the complete substrate. Basically the above is an overall functional test.

Although being an example of a professional microsystem, only a fraction of the above tests will be carried out in the production testing stage, as repair of the non-electrical parts of the microsystem is very difficult or impossible. It is further noted that the functionality of some of the mechanical parts is not explicitly tested. Any jamming of the micro fluid channels (e.g. by dust particles), or failing micropump valves will appear as failing flow sensor and pump operation.

5.   CONCLUSIONS AND RECOMMENDATIONS

In the past, many test methodologies and support hardware and software have been developed in the area of ICs, PCBs and systems. Many of the current techniques can also be applied to microsystems. Although developments in digital parts are well matured, still research effort is required with regard to high-speed behaviour in computer microsystems. The current lacks in analogue and mixed-signal IC and PCB testing also hamper the test development of mixed-signal microsystems. A mixed-signal Boundary-Scan approach, in combination with powerful on-board DSP, enabling BIST, is expected to be the major test approach in the future.

Still much work is required with regard to testing the sensor and actuator parts of microsystems. Some types lend themselves for BIST, while others require external stimuli and evaluation and even substrate assembly before testing. Non-electrical failures can sometimes be indirectly measured in electrical quantities, but often not. For sensors and actuators, mixed-signal boundary-scan techniques hold a great promise. It turns out that the test approaches used are to a great extend depend on the actual implementation and application area of microsystems, and still require a tremendous effort in research.

REFERENCES

Abramovici, M., Breuer, M. and Friedman, A. (1990) Digital Systems Testing and Testable Design, *IEEE Press*, ISBN 0-7803-1062-4.

Arabi, K., Kaminska, B. and Sunter, S. (1996) Testing Integrated Operational Amplifiers Based on Oscillation-Test Method, *International Mixed-Signal Test Workshop*, 227-232.

Beenker, F. (1986) Macro testing, Unifying IC and Board Test, *IEEE Design and Test of Computer Magazine*, 26-32.

Bleeker, H. (1992) Boundary-Scan Testing: A Practical Approach, *Kluwer*.

Bracho, S. (1997) Dynamic Current Testing Methods in Mixed-Signal Integrated Circuits, *Proc. ESPRIT "AMATIST" Open Workshop*, Enschede, 6.1-6.9.

Kawahito, S. (1995) Micro-Fluxgate Magnetic Sensing Elements using Closely Coupled Excitation and Pickup coils, *Transducers 1995, Sensors and Actuators, 290-A12*, 233-236.

Kerkhoff, H. and Tangelder, R. (1995) Testable Design and Testing of Integrated Microsystems, *University of Twente*.

Kerkhoff, H. et al. (1997) MCM-D Mixed-Signal Boundary-Scan Testing, *European Test Workshop*, Cagliary, 9.1-9.5.

Lammerink, T. et al. (1995) Intelligent gas-mixture flow sensor, *Sensors and Actuators, Elsevier*.

Mahoney, M. (1987) DSP-based Testing of Analogue and Mixed-Signal Circuits, *IEEE Computer Society Press*.

Ohletz, M. (1991) Hybrid Built-In Self-Test (HBIST) for Mixed Analogue/Digital Integrated Circuits, *Proc. of the European Test Conference*, Paris, 307-316.

Rajsuman, R. (1995) Iddq Testing for CMOS VLSI, *Artech House Inc.*, Norwood.

Tangelder, R., Diemel, G., Kerkhoff, H. (1997) Smart sensor system application: An Integrated Compass, *European Design & Test Conference*, Paris, 195-199.

Van den Berg, A. and Bergveld, P. (1996) Development of MicroTAS Concepts at the MESA Research Institute, *MicroTAS'96*, 9-15.

Van der Plassche, R. (1994) Integrated Analogue-to-Digital and Digital-to-Analogue Converters, *Kluwer Academic Publishers*.

Weyerer, M. and Goldemund, G. (1992) Testability of Electronic Circuits, *Prentice Hall*, Hempstead.

Wilkins, B. (1997) P1149.4 standard for a Mixed-Signal Test Bus, *IEEE*.

Wilkinson, J. (1997) Microsystems Technology and Applications, *Tutorial VLSI'97*.

CHAPTER 11

EMBEDDED CORE-BASED SYSTEM-ON-CHIP TEST STRATEGIES

YERVANT ZORIAN

LogicVision, Inc., 101 Metro Dr, Third Floor, San Jose, CA 95110, USA, Phone: +1-408-453-0146, Fax: +1-408-467-1186, e-mail: zorian@logicvision.com

Abstract: Chips containing reusable cores, i.e. pre-designed Intellectual Property (IP) blocks, have become an important part of IC-based systems. Using embedded cores enables the design of high-complexity systems-on-chip with densities as high as millions of gates on a single die. The increase in the use of pre-designed IP cores in system-chips adds to the complexity of test. To test system-chips adequately, test solutions need to be incorporated into individual cores and then the tests from individual cores need to be scheduled and assembled into a chip level test strategy. However with the increased usage of cores from multiple and diverse sources, it is essential to create standard mechanisms to make core test plug-and-play possible. This chapter presents in general the challenges of testing core-based system-chips and describes their corresponding test solutions. It concentrates on the common test requirements and strategies

Keywords: Testing Embedded Core, IEEE P1500 embedded core test standard, Intellectual Property Test, System-on-Chip Test

1.   INTRODUCTION

The need for faster and smaller products has recently driven the semiconductor industry to introduce a new generation of complex chips. Chips that allow the integration of a wide range of complex functions, which used to comprise a system, into a single die, called system-on-chip.

While meeting today's performance and miniaturization requirements by physically placing these functions into a single chip design, a number of new issues are faced. One such issue is the complexity of creating multi-million gate system-chips from scratch using conventional methods and design flow. Hence, the design community has created a new chip design paradigm based on design reuse. This is realized by embedding pre-designed functions into a single system-on-chip (SOC).

R. Reis et al. (eds.), Design of Systems on a Chip, 219–232.
© 2006 *Springer.*

That is designing-in specialized reusable cores, i.e. megacells, into complex chips in order to minimize the time-to-design while taking full advantage of today's semiconductor technology.

The need for specialized components in the new SOC design paradigm is analogous to the conventional system-on-board one, where PCBs (printed circuit boards) are populated with specialized standard IC components. Although designing with cores looks conceptually similar to designing systems-on-board, in fact, the system-on-chip is challenged by fundamental differences. One major difference is, instead of plugging in a number of stand-alone manufactured components into a PCB, a system-on-chip designer needs to embed cores that are not manufactured, and hence are untested. For systems organizations, this is not a totally new phenomenon. They have practiced design reuse for years by integrating simple macros and in-house cores into ASICs. However, these cores were not mixed and matched from multiple external sources. Today, because of the diversity and complexity of specialized system functions, it is not economical for a typical organization to create all the required intellectual property (IP) functions by itself. Hence, IP cores are being leveraged from multiple sources. The variety of sources today provides a diversity of trade-off and formats and opens the door for plug and play requirements in chip design. This paradigm shift in chip design is sometimes termed as "system level integration" or "clip-art era in semiconductor design". Dataquest expects that mainstream ASIC designers fill 90% of their silicon area with embedded cores, 40-60% of which using external cores and the rest from internally developed one, while leaving only 10% of the chip for the application specific part called User Defined Logic (UDL) [15].

The embedded cores today not only cover a wide range of system functions, but also contain an unprecedented range of technologies, from logic to memory to analog. Furthermore, they may come in hierarchical compositions too. For instance, a complex core may embed one or more simple cores. An example SOC design is shown in Figure 1. Moreover, they come in a wide range of hardware description levels. They spread from fully optimized layouts in GDSII format to widely flexible RTL codes. Embedded cores are categorized into three major types based on their hardware description level: soft, firm and hard [17]:

1. A soft core is a synthesizable RTL code or generic-library netlist description of a piece of logic.
2. A firm core is also in the form of synthesized RTL code or a generic-library netlist, but a designer has optimized the core's structure and topology for performance or size. A firm core, which is not routed, may be optimized with floorplanning and, sometimes, placement for a few targeted technologies. Firm cores are more flexible than hard cores and more design parameter predictable than soft cores [9].
3. A hard core exists in layout and is targeted for a specific technology. A hard core is optimized for one or more key parameters, such as speed, size, or power. A hard core comes as a placed and routed netlist, physical-layout file (often in industry-standard GDSII format), or a netlist/layout combination. Each type of core has different modeling and test requirements.

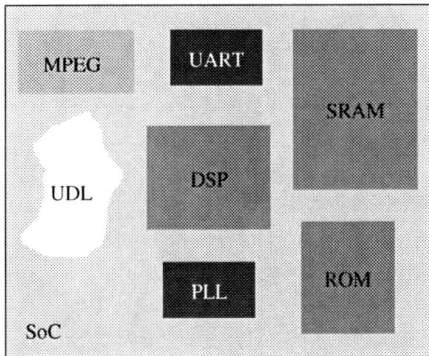

Figure 1. System-on-Chip consisting of hierarchical cores and User Defined Logic (UDL)

The three types of cores offer trade-offs. Soft cores leave much of the implementation to the designer, but are flexible and process-independent. Hard cores have been optimized for predictable performance, but lack flexibility. Firm cores offer a compromise between the two.

The emerging process of plug-and-play with embedded cores from diverse sources faces numerous challenges in the areas of system-on-chip design, integration and test. The list of challenges is typically headed by the complexity of test and diagnosis. This chapter analyzes the above challenges and discusses the current solutions to create a testable core-based system-chip. The chapter mainly covers the common needs of the industry and shed light on the current industry practices. In Section 2, the chapter discusses the testing issues in core-based system-on-chip. Section 3 describes the existing solutions to address core internal test, core peripheral access, and system-on-chip test integration. Finally, Section 4 concludes the chapter.

2.   ISSUES IN TESTING CORE-BASED SYSTEM CHIPS

In this section, the main testing issues in the new system-on-chip paradigm are analyzed and compared to the conventional system-on-board (SOB) ones. In addition to the major differences discussed in this section, we have to note that system-chips do also share all the testing issues of the traditional deep-submicron chips, such as defect/fault coverage, overall test cost and time-to-market. Although the design processes in both paradigms, i.e. in SOC and SOB, are conceptually analogous, the test processes are quite different. In the conventional SOB, manufacturing and test of a standard IC or a user defined ASIC take place first, before the PCB is assembled and tested; whereas in a core-based system, there is only a single manufacturing and test instance that takes place for the whole system-on-chip (SOC), see Figure 2. In this case, the individual cores and their surrounding User Defined Logic (UDL) are designed individually first. Next, they are all integrated into an

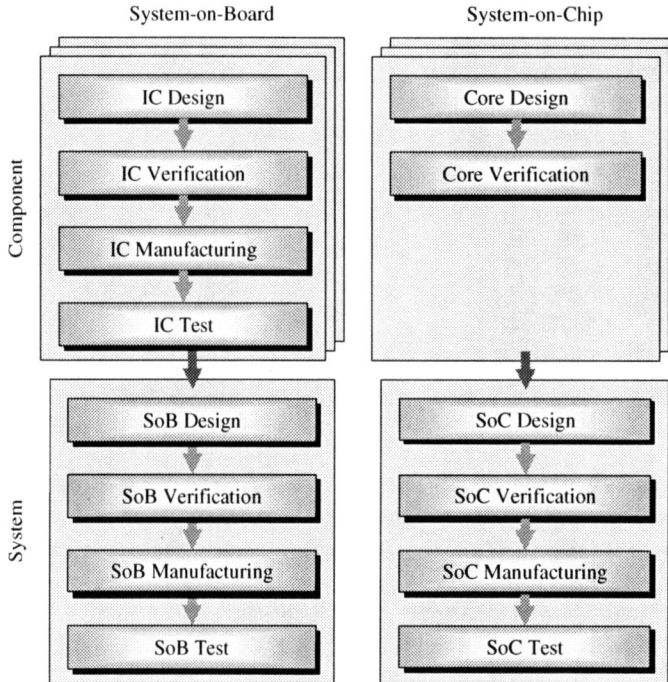

Figure 2. Multi-stage conventional System-on-Board Test versus single stage System-on-Chip Test

SOC. And only then, the composite manufacturing and test steps take place. See the comparison of the two processes in Figure 2.

2.1 Core Internal Test

The internal test of a core is typically composed of internal DFT structures (e.g. scan chains, test points), if any, and the required set of test patterns to be applied or captured on the core peripheries. The test patterns need to include data and protocol patterns. The data patterns contain the actual stimulus and response values; whereas the protocol patterns specify the test flow, i.e. how to apply and capture the data patterns (e.g. shift and apply sequence in scan protocol). Test protocol patterns are independent of the data patterns and need to access the core peripheries to execute a given test.

A core is typically the hardware description of today's standard ICs, e.g. DSP, RISC processor, DRAM. A given core is not tested individually as in standard ICs, but is instead tested as a part of the overall SOC (see Figure 2). However, the preparation of a core's internal test is typically done by the core creator. This is because the core integrator in most cases, except for soft cores, has very limited knowledge about the structural content of the adopted core and hence deals with it as a black box. Therefore he/she cannot prepare the necessary test for it, especially if a core is a hard one or is an encrypted Intellectual Property block. This requires that the core creator prepares the internal test capability and delivers it with the core, i.e. the DFT structures and the test patterns.

Another major issue for the core creator is to determine the internal test requirements of the core without knowing the target process and application. For instance, which test method needs to be adopted (e.g. BIST, scan, Iddq, functional test), what type of fault(s) (e.g. static, dynamic, parametric) and what level of fault coverage is desired. In SOB, a chip's ppm figures are known prior to the PCB assembly. Hence, the test method and desired fault coverage are predetermined accordingly. But in SOC, a core creator might not know the target process and the desired quality level. Hence, the provided quality level might or might not be adequate. If the coverage is too low, the quality level of the SOC is put at risk, and if it is too high the test cost may become high (e.g. test time, performance, area, power). Furthermore, different processes have different defect type distributions and levels of defect. Furthermore, in system testing, the failure mechanisms differ from those in production and the test time and power requirements (e.g. for BIST) may be different. Hence, a core should be well characterized with respect to the test and fault coverage. Ideally, the internal test should be a composite one containing modular sections with different characteristics. These modular sections can be switched in or out according to the target process and application. Different internal core test approaches are discussed in Section 3. Finally, the core internal test prepared by a core creator needs to be adequately described, ported and ready for plug and play, i.e. for interoperability, with the SOC test. For an internal test to accompany its corresponding core and be interoperable, it needs to be described in a commonly accepted, i.e. standard, format. Such a standard format is developed by the IEEE P1500 [21].

2.2   Core Test Access

Another key difference between SOB and SOC is the accessibility of component peripheries, i.e. accessing primary input/outputs of chips and cores, respectively. With SOB, direct physical access to chip peripheries, i.e. pins, is typically available to use probing during manufacturing test; whereas for a core, which is often deeply embedded in an SOC, direct physical access to its peripheries is not available by default, hence, an electronic access mechanism is required, i.e. additional wiring and logic to connect core peripheries to test resources. To establish an electronic access mechanism, two basic elements need to be incorporated into the design: a set of access paths and a mode control network to enable/disable the desired access paths:

1. Access Path: An access path is meant to connect a core periphery (e.g. input, output, or bidirectional) to the assigned test resource block. Figure 3 demonstrates three access paths connecting core inputs P1, P2, and P3 to three possible test resources T1, T2, T3. The first test resource is based off-chip, e.g. functional patterns applied from an Automatic Test Equipment. T2 is connected to an on-chip test response. It may be a BIST Test Pattern Generator [1], a functional

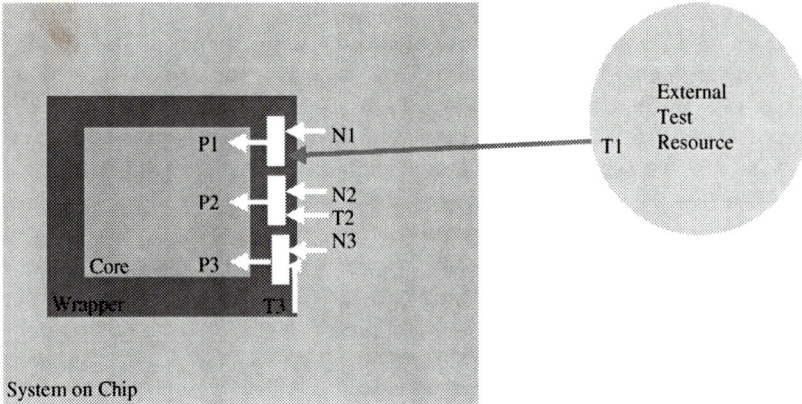

Figure 3. Embedded Core Accessibility

resource [2], a soft grid matrix [4], etc. Finally, T3 is a connected to a test resource internal to the Peripheral Access block, such as a scan source from a neighboring cell supplying serialized functional or structural patterns. N1, N2, and N3 are normal mode inputs to the Embedded Core. The Peripheral Access block contains the necessary DFT structure for all core peripheries to switch between normal and test modes. Various types of Peripheral Access blocks, which are compatible with internal test methods and optimized for area and performance constraints, are summarized in Section 3.2.

2. Mode Control: The peripheral access paths need to be controlled either from on-chip or off-chip controllers. Mode control, as shown in Figure 3, is necessary for a given peripheral access box to select (i.e. enable and disable) the paths according to the desired mode. In addition to the normal and core internal test modes, a System-on-chip often requires several other test related modes. One such mode is the core external test. The role of an external test is to test the interconnect faults (static or dynamic) between the cores in a System-on-chip (e.g. interconnect faults between the DSP core and the Embedded DRAM in Figure 4). Also, core external test is needed to participate in the UDL test, by providing controllability and observability to the UDL peripheries through the Peripheral Access block of a neighboring core (e.g. the access path of the DSP core in Figure 4). In addition to applying and capturing test patterns during internal and external test modes, peripheral access can be utilized for core isolation. Typically, a core needs to be isolated from its surroundings in certain test modes. Core isolation is often required on either the input or the output side. On the input side, it puts the core in a safe-state by protecting the core under test from external interferences (from preceding cores or UDL). On the output side, it protects the inputs of the superseding blocks (cores or UDL) from undesired values (e.g. random patterns applied to tri-state buffers creating bus conflicts). This is done by controlling the core outputs and putting them in tri-state mode

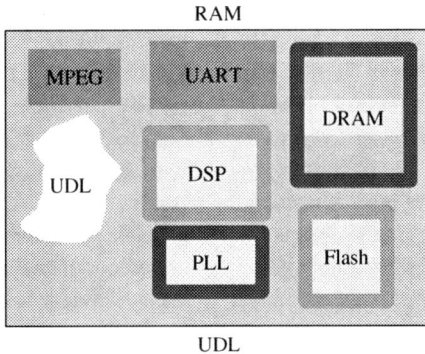

Figure 4. Embedded Cores with Peripheral Access

(core in Hi-Z state). Sometimes, other core isolation modes are also required to put a core in a stand-by mode, or low power dissipation mode, or ByPass mode, or finally to control the current supply of a core, if Iddq testing is used in the System-on-chip. The Peripheral Access block being an interface block between a core and its surrounding logic (UDL or other cores) provides the necessary support for the above access and isolation modes. It allows these modes to be selected through the mode control protocols which control the switching capabilities of the Peripheral Access block. In Section 3.2, this chapter summarizes popular DFT techniques to provide peripheral access for embedded cores.

Since cores are imported from diverse sources, a common set of Peripheral Access mechanisms and core test control mechanisms are required, as addressed by the IEEE P1500 [21].

2.3   System-on-Chip Test and Diagnosis

One of the major challenges in the SOC realization process is the integration and coordination of the on-chip test and diagnosis capabilities.

If compared to SOB, the system-chip test requirements are far more complex than the PCB assembly test, which simply consists of interconnect and pin toggling tests. The SOC test, as in Figure 2, is a single composite test. This test is comprised of the individual core tests of each core, the UDL test, and the test of their interconnects. As discussed earlier, each individual core or UDL test may involve surrounding components. Certain peripheral constraints (e.g. safe-mode, low power mode, bypass mode) are often required. This requires corresponding access and isolation modes. Composite test requires adequate test scheduling. Test scheduling is needed to meet a number of chip level requirements, such as total test time, power dissipation, area overhead, etc. Also, test scheduling is necessary to run intra-core and inter-core tests in certain order not to impact the initialization and final contents of individual cores. With the above scheduling constraints the schedule of the composite SOC test is created. In addition to the test integration and interdependence issues, the SOC.

The mode control of individual Peripheral Access blocks contributes to the execution of the composite test. A chip level mode control DFT network is required to run the composite test and apply/capture the necessary on-chip and off-chip test patterns (i.e. data patterns and protocols). In addition to the composite SOC test, which is meant to be used as manufacturing test, other tests, debug and diagnosis modes are often required. For instance, test modes are required for field testing and debug modes for the SOC development. Furthermore, diagnostic modes are often very critical to the production of SOC. They are required to identify failed cores or in some cases to determine the internal failures of each core. Fault isolation requirements may impact the adopted internal core test method. In certain cases, the identification of a failed core may not be possible, due to the fact that the cores pass their individual intra-core tests, but the composite test detects an inter-core fault, such as an interconnect fault or a delay fault. To summarize, there are three major sets of issues in testing core-based systems: creating adequate and portable core internal tests; providing core accessibility; and creating an integrated test and its control mechanism for the overall system-on-chip. The following section presents a taxonomy of current solutions that address these three sets of issues.

3.   TAXONOMY FOR CORE-BASED TESTING

In the conventional SOB paradigm, the roles and responsibilities for providing IC test and PCB test are clear. But the same does not apply for core-based systems. Here, the test preparation function crosses the boundaries, from core creation phase, to SOC integration phase, to silicon fabrication phase. In each phase a portion of the overall test is completed. Typically, the core internal test is prepared at the core creation phase. The extent of the core preparation in this phase depends on the type of the core, i.e. soft, firm, or hard. In some cases, full DFT needs to be incorporated in the core creation phase, and in others only test scripts and benches are generated in this phase. The Peripheral Access mechanism for individual cores is determined and incorporated with at the core creation phase or at the system-on-chip integration phases. The UDL test, the assembly of the composite test and the creation of the test control network are performed at the SOC integration phase. All above test features are utilized at the fabrication phase. The manufacturing test is executed based on the provided capabilities. Also, at the fabrication phase, the DFT circuits for debug and diagnosis are used. The test information transfer from one phase to the other, i.e. from one player to the other, requires standardization in order to guarantee interoperability and ease of plug-and-play in the creation of system-chips [21].

Based on the three sets of requirements presented in the previous section, here we categorize the technical solutions that address these requirements, and discuss the necessary activities at their corresponding phases.

3.1   Embedded Core Internal Test

In this subsection, we decouple the core internal test from testing a systems-on-chip, and we concentrate on the first, leaving the second for the following two subsections.

Incorporating an adequate internal test method for an embedded core, such as BIST, ATPG/scan, or functional patterns, is not very different to selecting a test method for a conventional ASIC. The problem here is that the core integrator often cannot make the test method selection independently of the core creator. Because the function of preparing an internal test for an embedded core is distributed between the core creator and user. The extent of the involvements by each depends on the hardware description level of the core. If the core is mergeable with the User Defined Logic (UDL), than the core integrator has more flexibility in identifying and implementing a core test method; whereas if the core is non-mergeable then the core creator has a much larger role in determining and incorporating the core test method(s).

Moreover, depending on the requirements, a core test may need to contain design-for-testability features that guarantee fault coverage, and provide low impact on area and timing, and often a certain degree of diagnosis.

3.1.1   Mergeable cores

A core is defined mergeable if it is a soft or firm (RTL or gate level) and can be combined with its surrounding random logic via standard flow without the need to create an isolation layer (i.e. Peripheral Access) between itself and the UDL around it. In this case, the internal test method is not pre-defined during the core creation stage. The core integrator merges the core with the surrounding logic and then applies the same test method (e.g. scan, ATPG, BIST) on the merged logic, i.e. the core and the UDL [8].

Figure 5 shows the RTL level MPEG core of Figure 4 merged with the neighboring UDL to create an extended UDL for test purposes. In the creation phase, the extent of test preparation for mergeable cores is limited to passing audits, generating DFT scripts, running fault simulation with the selected test method to estimate the fault coverage and the vector counts.

Describing the above test related information by the core creator and porting it to the integrator in a standard format would be very beneficial.

3.1.2   Non-mergeable cores

A core is considered non-mergeable if it cannot be absorbed by its surrounding logic. Hence, it remains as a distinct entity. This happens because:
1. The core hardware description level is different from the one used in the surrounding logic. For example, the DSP core shown in Figure 5 cannot be merged with the surrounding UDL; Because it is a layout level core.
2. The core uses a different technology, such as SRAM, Flash or analog.

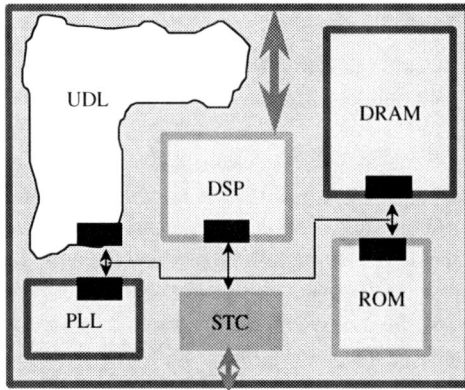

Figure 5. System-on-Chip with Core Internal Test, Peripheral Access and Core Test Interface

3. The core needs to remain isolated for IP protection. In this case, even an RTL core can not be merged with its surrounding UDL due to its encryption. In the case of non-mergeable cores, typically the internal core test is incorporated mainly during the core creation phase and it is based on the target technology. In addition, the history of the core may have an impact on the internal test method. For instance, almost all memories today tend to use BIST, hence providers of memory core generators typically incorporate the BIST wrappers and certain BIST control signals into the memory core design [20].

With non-mergeable cores, most of the test preparation is performed at the core creation phase. That would include all the necessary steps to incorporate the test blocks associated with the given test method (e.g. test points for BIST, scan chains, etc.), generate test patterns (e.g. ATPG patterns) and provide fault coverage data. The core test method is assembled with the rest of the system-on-chip test during the core integration by the system designer(s), and finally the test is executed during the chip fabrication process.

The basic core test methods used for mergeable and non-mergeable cores are typically the same as the ones used in conventional ASICs today. They either depend on external resources or are self-contained with minimal need for external test patterns.

The tests that require external patterns are based on functional test patterns or structural (ATPG generated) patterns. The functional patterns are created for design verification purposes and typically do not provide high fault coverage, especially for large cores. Various cores today do provide functional vectors with the delivery of a core. Examples of such cases are RISC cores, some memory cores, and analog cores.

ATPG generated patterns are based on structural testing, hence provide measurable fault coverage. But they do not typically achieve high fault coverage unless coupled with a DFT approach, such as scan. In such a case, the scan chain is pre-synthesized during the core creation stage of a non-mergeable core and during the integration stage for mergeable cores. ATPG patterns are often generated by the core user for un-encrypted cores. The ATPG patterns mainly target stuck-at faults, and sometimes performance (delay) and Iddq faults are also targeted. Numerous processor and peripheral cores are delivered today with their scan chains and ATPG vector sets.

The problem with external test pattern based methods is that they typically require large test volumes and need complex test access protocols to get stimulus to cores, to receive responses from cores and to set up the necessary control to execute the test on a core. An alternative scheme that minimizes such requirements is Built-In Self-Test. BIST generates and evaluates the test patterns on-chip, hence does not require porting of test vectors from the core creator to the integrator and then to the fabricator. BIST is an autonomous testing method and is considered ideal for core-based systems.

In addition to providing the on-chip test resources, BIST simplifies peripheral access requirements, by simplifying the interface and allowing re-use of the core as a self-contained block. Today's BIST schemes are mature enough in terms of providing very high fault coverage. They also enhance the diagnoseability, while allowing IP protection.

The core test scheme has to provide modular interface to allow integration of hierarchical test control scheme and allow potential sharing of test resources at higher levels.

Even though the increased presence of BIST in embedded cores is a fact, it seems not to be the only solution at least for sometime. With a number of test methods used for core internal test, there seems to be no possibility for standardizing an internal test method. However, the delivery mechanism of the core test from core providers to core integrators need to be standardized.

This may happen through the interface mechanism between the core and the system-on-chip. The next subsection discusses the hardware interface between a core and its surrounding UDL. The common needs in this domain are obvious and a common mechanism definitely helps create interoperability of cores from diverse sources.

3.2   Embedded Core Peripheral Access

Typically, cores are deeply embedded in a system-on-chip. In order to apply or exercise the test of a given core, an access to its peripheries is often necessary. Such an access un-embeds the core, hence providing full controllability and observability to its peripheries. In addition to accessing, this provides effective isolation for IP protection and for putting the core and the surrounding logic in safe modes. A peripheral access around a core fully contributes to the test of the surrounding UDL, by providing observability to the core inputs and controllability to the core outputs.

Being between the core and the UDL, a peripheral access is either implemented during core creation or later during system integration. Traditionally, there are three major peripheral access techniques.

3.2.1   Parallel direct access

The technique is based on parallel access to the core I/Os by either using dedicated chip level terminals or by sharing chip terminals while multiplexing and demultiplexing them with other signals. This approach, which is termed as Star approach, is popular when the number of cores is very limited and the number of chip pins is more than the core pins. But with the continuous increase in the ratio of core terminals to chip terminals, this peripheral access technique is reaching its limits. A number of proprietary techniques exist today to map core peripheries to chip primary input/outputs, even if the later is smaller in number [13].

A major disadvantage in this access mechanism is its very high routing cost. Also, special attention need to be applied if at-speed testing is required due to problems with access path delays, skews, etc.

3.2.2   Serial scan access

The second technique for peripheral access is based on serialization of test patterns. A scan chain around the core allows indirect but full access to all the I/Os of a core. This is also called ring access mechanism. The rings can be internal or external to the core and can be implemented by the core creator or the core integrator. In addition to accessing the core, such a ring also provides effective isolation from the surrounding UDL of the core, which may be necessary to run different tests in the core and its surrounding logic. The complexity and hardware requirements for the ring approach are acceptable for today's system-on-chips and it remains independent of pin limitations. The use of the Boundary-Scan chain of IEEE 1149.1 as a serial access mechanism has been used by several core providers such as ARM and TI [18].

The routing cost of this serial approach is far less than the previous one.

Also, the serial approach allows effective isolation and UDL test if the scan chain is used in its external test mode. The two negative aspects of this approach are the test time increase, due to serialization, and the arbitrary signal switching during scan (control and clock signals need to hold scan cell output stable).

Several DFT approaches were introduced to minimize the hardware cost of the serial scan ring around a core, by using existing functional flip-flops, mixing internal and external FFs, or using compaction and expansion [6] [16].

In realistic examples often a combination of serial and parallel access cells are used for a given core, where the data patterns are applied through the serial ring, whereas the clocks and control signals are asserted via parallel access.

3.2.3   Functional access

The third technique for peripheral access is based on using the surrounding logic to propagate and justify the necessary test patterns [3] This can be either based on existing normal mode logic put in transparency for core test purposes [2], or be based on using existing DFT mode, such as scan chains in the surrounding logic. Even though, the cost of extra hardware is very low, this approach becomes complicated if the surrounding logic has deep functional paths and requires sophisticated protocols for test pattern translation.

The timing impact of a peripheral access mechanism is critical. The path delay created by peripheral access logic and the skews introduced in the access path need to meet the requirements of core test timing constraints.

The ring-based technique remains the most realistic option for peripheral access. The ring approach provides access to run internal BIST and internal scan for each core. It also helps testing the surrounding logic using the ring registers.

3.3   Test Integration for Core-based Systems

In addition to peripheral access, the system-on-chip requires an integrated test solution which connects the test facilities of each core and the DFT of the surrounding logic under one test control mechanism. The chip level control circuitry needs to be hooked to the Boundary-Scan port of the chip. Hence, providing a standard access mechanism via the Boundary-Scan TAP.

The test session for the system-on-chip has to be composed of intra-core testing to guarantee that each core has been manufactured correctly, and then an inter-core test is performed throughout the surrounding logic. The test execution schedule has to take into account several variables such as test time, power dissipation, noise level during test, and area optimization [20].

4.   CONCLUSIONS

This chapter presents a number of testing issues facing core-based system-chips. It discusses the current test strategies and identifies a set of common industry requirements to allow for interoperability of cores from diverse sources.

REFERENCES

1. V.D. Agrawal, C.J. Lin, P.W. Rutkowski, S. Wu, Y. Zorian, "Built-In Self-Test for Digital Integrated Circuits", AT&T Technical Journal, Vol. 73, No. 2, p. 30, March 1994.
2. F. Beenker, B. Bennetts, L. Thijssen, "Testability Concepts for Digital ICs – Macro Test Approach" Volume 3 of Frontiers in Electronics Testing. Kluwer Academic Publishers, Boston, 1995.
3. B. Bennetts, "A Design Strategy for System-on-a-chip testing", Electronic Products, June 1997, pp. 57-59.
4. S. Bhatia, T. Gheewala, P. Varma, "A Unifying Methodology for Intellectual Property and Custom Logic Testing", Proc. of International Test Conference, Oct. 1996, pp. 639-648.
5. R. Chandramouli, S. Pateras, "Testing Systems on a Chip", IEEE Spectrum, pp. 42-47, Nov. 1996.
6. K. De, "Test Methodology for Embedded Cores which Protects Intellectual Property", VTS 97, pp. 2-9.
7. R.K. Gupta, Y. Zorian, "Introduction to Core-based System Design", IEEE Design & Test of Computers, Vol. 14, No. 4, Oct.-Dec. 1997.
8. S.G. Hemmady, T.L. Anderson, Y. Zorian, "Verification and Testing of Embedded Cores", Proc. Design SuperCon, On-Chip Design, S122-1-19, Jan. 1997.
9. M. Hunt, J.A. Rowson, "Blocking in a System on a Chip", IEEE Spectrum, pp. 35-41, Nov. 1996.

10. B.K. Koenemann, K. Wagner, "Test Sockets: A Test Framework for System-On-Chip Designs", IEEE TTTC Embedded Core Test TAC Study Group, April 1997.

11. J. Lipman, "The Hard Facts about Soft Cores", EDN, pp. 35, Sept. 2, 1996.

12. G. Maston, "Structuring STIL for Scan Test Generation based on STIL", Porc. Int'l Test Conference, 1997.

13. J. Monzel, "Low Cost Testing of System-On-Chip", Proc. Digest of 1st IEEE International TECS, Nov. 1997.

14. B.T. Murray, J.P. Hayes, "Testing IC s: Getting to the Core of the Problem", IEEE Computer, Vol. 29, No. 11, pp. 32-38, Nov. 1996.

15. G. Smith, "Test and System Level Integration", IEEE Design & Test of Computers, Vol. 14, No 4, Oct.-Dec. 1997.

16. N. A. Touba, B. Pouya, "Testing Embedded Cores Using Partial Isolation Rings", Proc. of the 15th IEEE VLSI Test Symposium, pp. 10-15, April, 1997.

17. "Virtual Socket Interface Architectural Document", VSI Alliance, Nov. 1996.

18. L. Whetsel, "An IEEE 1149.1 based Test Access Architecture for ICs with Embedded IP Cores", Porc. Int'l Test Conference, Nov. 1997.

19. P. Wohl, V.T. Williston and J. Waicukauski, "A Unified Interface for Scan Test Generation based on STIL", Proc. Int'l Test Conference, Nov. 1997.

20. Y. Zorian, "A Distributed BIST Control Scheme for Complex VLSI Devices", Proc. of the 11th IEEE VLSI Test Symposium, p. 6-11, April 1993.

21. Y. Zorian, "Test Requirements for Embedded Core-based Systems and IEEE P1500", Proc. of IEEE International Test Conference, Nov. 1997.

22. Y. Zorian, "Testing Embedded Core-based System-Chips", Proc. of IEEE International Test Conference, Oct. 1998.

INDEX OF AUTHORS

233